T0296069

CAMBRIDGE MONOGRAPHS IN
EXPERIMENTAL BIOLOGY
No. 6

THE PHYSIOLOGY OF
REPRODUCTION IN FUNGI

THE SERIES

1   V. B. WIGGLESWORTH: The Physiology of Insect Metamorphosis.
2   G. H. BEALE: The Genetics of *Paramecium aurelia*.
3   G. V. T. MATTHEWS: Bird Navigation.
4   A. D. LEES: The Physiology of Diapause in Arthropods.
5   E. B. EDNEY: The Water Relations of Terrestrial Arthropods.
6   LILIAN E. HAWKER: The Physiology of Reproduction in Fungi.
7   R. A. BEATTY: Parthenogenesis and Polyploidy in Mammalian Development.

Other volumes in preparation

THE PHYSIOLOGY
OF REPRODUCTION
IN FUNGI

BY

LILIAN E. HAWKER, D.Sc.

Reader in Mycology in the University of Bristol

CAMBRIDGE

AT THE UNIVERSITY PRESS

1957

CAMBRIDGE
UNIVERSITY PRESS

University Printing House, Cambridge CB2 8BS, United Kingdom

Cambridge University Press is part of the University of Cambridge.

It furthers the University's mission by disseminating knowledge in the pursuit of education, learning and research at the highest international levels of excellence.

www.cambridge.org
Information on this title: www.cambridge.org/9781316509883

First published 1957
First paperback edition 2015

A catalogue record for this publication is available from the British Library

ISBN 978-1-316-50988-3 Paperback

CONTENTS

ACKNOWLEDGEMENTS   *page* viii

I INTRODUCTION   I

   Types of reproduction among fungi: (*a*) vegetative; (*b*) reproduction by spores. Functions of fungal spores.

2 THE GROWTH OF SPORES AND OF SPORE-
   BEARING STRUCTURES   7

   Spores. Sporangiophores and sporangia of *Phycomyces*. Complex fruit-bodies. Sclerotia.

3 THE PHYSIOLOGY OF VEGETATIVE REPRO-
   DUCTION   16

   Oidia. Chlamydospores. Sclerotia.

4 THE EFFECT OF ENVIRONMENT ON
   SPORULATION   24

   Temperature. Water. Aeration. Hydrogen-ion concentration
   of the substrate. Various types of injury (toxins, mechanical
   injury, barriers to growth). Light. Gravity. Interaction of
   environmental factors.

5 THE EFFECT OF NUTRITION ON SPORULATION   48

   Concentration of food material. The nature of the food supply:
   (*a*) source of carbon; (*b*) nitrogen source and carbon-nitrogen
   ratio; (*c*) phosphorus, potassium, magnesium and sulphur;
   (*d*) calcium; (*e*) trace elements; (*f*) growth-substances.
   Metabolism and the reproductive phase.

6 THE PHYSIOLOGY OF SEX   76

   Heterothallism. Sex hormones and specific 'fruiting' substances. The determination of sex. Loss of sexuality.

7 REPRODUCTION IN THE NATURAL HABITAT   100

   Summary of conclusions from pure culture studies. The
   natural habitat: (*a*) production of fruit-bodies by the larger
   fungi; (*b*) plant pathogens.

REFERENCES   107

INDEX   124

ACKNOWLEDGEMENTS

My thanks are due to Professor W. Brown, F.R.S., who introduced me to the study of the physiology of fungi, to Dr S. D. Garrett, Botany School, Cambridge, for kindly reading the manuscript and making a number of helpful comments, to a number of colleagues at Bristol University and elsewhere, for information and comments on various specific points, and to Dr J. Fraymouth for assistance in the preparation of the index.

BRISTOL   L. E. H.
January 1956

CHAPTER I

INTRODUCTION

THE fungi are remarkable for diversity of both form and function. Their ability to break down and to synthesize complex substances largely accounts for their success in colonizing widely different habitats, and has made them of the greatest economic importance as parasites, as saprophytes in the soil and on varied commercial products, and as the producers of antibiotics, organic acids and other substances useful to man. Even more striking is the wide range of structure and life history within the group. The vegetative part of the fungal thallus is either unicellular, as in some primitive and a few degenerate forms, or, more commonly, filamentous. These fungal filaments, which are known as hyphae, usually branch freely and together form the mycelium or vegetative thallus of the organism. In the Lower Fungi (Phycomycetes), the actively growing vegetative hyphae are aseptate and coenocytic; septa form only to cut off dead or injured parts of the hypha, or in connexion with reproduction. Such a system is not suitable for the building-up of complex structures and hence reproductive bodies in this group are usually simple. In the Higher Fungi (Ascomycetes, Basidiomycetes and Fungi Imperfecti) the hyphae are regularly septate from an early stage and, moreover, have a strong tendency to anastomose. Among these fungi, complex structures may arise through interweaving and anastomosis of the hyphae.

While the factors controlling the onset of the reproductive phase are of great intrinsic interest in all organisms, in the fungi they are also of great economic importance, since it is through their reproductive bodies that the organisms are able to spread and colonize new substrata, and often are able to survive temporarily unfavourable conditions. A knowledge of the factors inducing reproduction may thus often solve the problem of the control of parasitic and other harmful fungi. Moreover, since the classification of fungi rightly rests largely on the nature of their reproductive bodies, the physiology of reproduction is of importance also to the taxonomist.

I

When a reproductive unit, such as a piece of mycelium or a spore or group of spores, becomes established on a suitable substratum, as when a culture plate is inoculated, it usually produces a colony which remains in the vegetative phase for some time. Sooner or later, however, with most fungi, provided that environmental conditions are favourable, specific reproductive bodies begin to form.

TYPES OF REPRODUCTION AMONG FUNGI

(a) *Vegetative.* The simplest type of reproduction in fungi is the rounding-off and ultimate separation of the vegetative cells. Many fungi, such as species of *Mucor* or *Endomyces*, produce these oidia in sugary media or with increasing age. In some fungi, or under different conditions, thick-walled cells with dense contents are produced, such as the 'Dauerzellen' of the unicellular yeasts or the chlamydospores of certain filamentous moulds. These are essentially small parts of the vegetative thallus adapted to survive periods of drought, of extremes of temperature or of exposure to solutions of high osmotic pressure, which would be lethal to the ordinary thin-walled vegetative cell. Little is known of the exact factors inducing the formation of chlamydospores, but it is generally assumed that these spores are formed in response to the onset of unfavourable conditions. They are produced by relatively few fungi. The gemmae of the water moulds may sometimes be developed from vegetative cells but are commonly formed from unfertilized oogonia.

Larger vegetative reproductive bodies, known as sclerotia, are produced by a number of fungi and particularly by plant parasites inhabiting the soil. Sclerotia consist of closely interwoven hyphae, which often lose their original form so that the individual cells become globose, and are packed with reserve food substances. In some sclerotia the outer layer (or layers) of cells has become a pseudosclerenchyma with thick, often black, cell walls. Sclerotia are formed in a number of ways (Townsend and Willetts, 1954) from single hyphae or from groups of branches or hyphal strands, but their function is always that of surviving temporary food shortage, desiccation or other adverse conditions. On the return to more favourable conditions they may germinate directly by giving rise to a mass of ordinary vegetative hyphae or, as in species of *Sclerotinia* and some other genera, they may give rise to complex spore-bearing fructifications.

(b) *Reproduction by spores.* The most usual form of reproduction by fungi is, however, sporulation. Spores are minute separable bodies with a special form characteristic of the particular species. There are many different kinds of spore and the same fungus may produce several different types at different stages in its life history or in response to different environmental conditions.

The commonest type of spore produced by fungi is asexual: the zoospore or non-motile sporangiospore in the simpler, more primitive species, and the conidium in the Higher Fungi. These are often produced in large numbers and permit the rapid spread of the fungus under favourable conditions. While they are often slightly more tolerant than the mycelium, of adverse conditions, such as desiccation, they are not usually resistant cells as their walls are often no thicker than those of the hyphae. Asexual spores typically lack vacuoles and their cytoplasm is of a relatively concentrated nature. They are usually capable of germination as soon as they are shed but rapidly lose their viability if they do not fall on to a suitable substratum. With many fungi production of asexual spores goes on continuously or in successive waves whenever, and as long as, conditions are suitable. With many plant parasites this ceases with the approach of winter. It is not known in the majority of cases whether this inhibition is due to lower temperature or to the changed condition of the host plant itself, that is, to nutritional factors. Many fungi either do not produce any other type of spore or are not known to do so. These are included in the arbitrary group of the Fungi Imperfecti.

The majority of fungi, however, also produce a second type of spore following a nuclear fusion. In the simpler fungi, including most of the Phycomycetes, this nuclear fusion is clearly of a sexual nature and the resulting spore is a sexually produced zygote. In the Higher Fungi the fusion is not usually a clearly sexual one but takes place between two nuclei derived from mycelia of different strains or even from the same mycelium. These nuclei may pair, but fusion is often delayed. Thus spore formation does not, in many of these fungi, follow immediately after nuclear pairing but takes place after an interval during which a secondary, binucleate, phase, such as the ascogenous hyphae of the Ascomycetes or the secondary mycelium of the

3

Basidiomycetes, may continue for some time. Eventually nuclear fusion takes place and spore formation follows. Such spores, though they are associated with a nuclear fusion and though they are distinct from the asexual spores, cannot accurately be described as 'sexual' spores. Hence the rather unsatisfactory terms, 'perfect stage' and 'perfect spores', have been introduced to distinguish them from the asexual or 'imperfect' stage and spores.

In many Ascomycetes and Basidiomycetes the typical ascospores or basidiospores may be produced in large quantities in or on a complex fruit-body or sporophore. The hyphae forming these bodies may show considerable differentiation into layers of different function, such as the outer protective layer or periderm, a nutritive layer and the spore-bearing layer or hymenium. Some of the large perennial fruit-bodies of such fungi as *Fomes* and *Ganoderma* are exceedingly complex and of a hard woody nature.

In considering the physiology of reproduction in fungi it is obvious that the factors inducing the formation of these widely different types of reproductive body are likely to show wide differences and to be distinct from those favouring maximum production of mycelium. Further, in the development of the complex sporophores of the Higher Fungi, there is evidence that the factors inducing the initiation of these are not necessarily the same as those leading to their maturation and to the ultimate production of viable spores.

The fact that young colonies pass through a purely vegetative phase before the onset of reproduction, has already been pointed out (p. 2). The length of this initial period of vegetative growth depends partly on the environment but also largely on the age and genetic constitution of the culture. No manipulation of the environment is able to reduce this phase to less than a particular minimum value characteristic of the species or strain of fungus concerned. One must assume that reproduction does not take place until the conditions *inside* the mycelium are suitable. Such an internal condition may be influenced by the environment, but is not entirely dependent upon it. Very little is known of the nature of these internal factors inducing reproduction, but if this problem could be solved it would be a great step forward towards the understanding of the physiology of reproduction.

4

Even the simplest form of reproductive body is a relatively complex structure, and is certainly the result of a chain of chemical processes controlled by a large number of enzymes, all of which must either be synthesized by the fungus or acquired ready-made from the environment. Several investigators have shown (see pp. 67, 81) that reproduction in a number of fungi is preceded by a period of intense respiratory activity. It is not yet clear whether reproduction is the result of qualitative or of quantitative changes in metabolism, but in either case the process is certainly exceedingly complex. Environmental factors will thus influence reproduction if they act to modify or inhibit any stage in this long chain of events. It is not therefore surprising if apparently contradictory reports occur, since the final effect of a particular factor depends upon the particular critical stage of the reaction chain which is sensitive to it. The interplay of factors and the occasional partial replacement of one factor by another may also be better understood if it is realized that there are more processes than one involved in the metabolism of reproduction and that each process may be influenced by various factors. Some information on the mechanism by which certain nutritional and other factors may influence reproduction has been obtained by the study of artificially produced mutants of various fungi which are often more exacting in their requirements than the parent or wild type.

While the exact effect of environment must be worked out for each species or strain, and for each type of reproduction individually, it is yet possible to generalize to some extent about the factors usually favouring reproduction.

FUNCTIONS OF FUNGAL SPORES

Fungal spores have many functions. As already pointed out, the asexual spores are the main means of dispersal of certain fungi to new areas. Their simple structure and the fact that they are formed in large numbers make them an efficient means of such spread. The same function of dispersal and colonization of new substrata is also found in many of the so-called 'perfect' spores of the Higher Fungi. Ascospores and basidiospores are often produced in large numbers and may be light enough for efficient wind dispersal. Many of the Lower Fungi, however, and some of the higher ones, produce thick-walled resting spores

5

capable of withstanding periods of drought or of extremes of temperature. Some of these spores, such as the oospores of many Oomycetes, the zygospores of the Mucorales and the teleutospores of some rusts, are incapable of germination until they have been subjected to periods of adverse conditions such as intense cold or desiccation. In those Higher Fungi in which ascospores or basidiospores are borne on, or in, complex fruit-bodies the resistance to adverse conditions depends rather on the protective layers of the fruit-body than on the structure of the spores themselves. Not only do these highly developed fruit-bodies protect the developing spores, but they are also often specially adapted for the efficient dispersal of the mature spores. It is significant that few of the higher Hymenomycetes and Gasteromycetes produce conidia to any extent, dispersal being efficiently achieved by the basidiospores.

In addition to resistance and dispersal, the 'perfect' spore stage permits the rearrangement of characters through hybridization. This is often of practical significance in agriculture, as in the well-known example of the origin, on the barberry, of new races of *Puccinia graminis* which may include forms able to attack an extended range of wheat varieties. Segregation of characters in fungi is not, however, entirely dependent on a nuclear fusion followed by a reduction division, since, owing to the strong tendency for anastomosis and consequent nuclear migration between allied strains, the nuclear content of the resulting complex mycelium is of mixed origin and may give rise to new combinations of characters (Hansen, 1938).

The significance of spores in the biology of the fungus and in man's attempts to control fungi is thus clear and the study of the physiology of the initiation and maturation of spores and spore-bearing structures is of great importance. While recent work has produced several detailed studies of spore production in particular species, the subject is one of almost unlimited opportunity.

The reproductive cycle is not complete until the spores have been discharged—often through the action of complex mechanisms—have alighted on a suitable substratum and have germinated and established a new mycelium. All these phases are influenced by the environment. They are outside the scope of the present account and have been reviewed elsewhere (Ingold, 1939, 1953; Gottlieb, 1950; Hawker, 1950; Gregory, 1952).

6

CHAPTER 2

THE GROWTH OF SPORES AND OF SPORE-BEARING STRUCTURES

THE physiology of sporulation in fungi can be well understood only after a consideration of the actual morphological changes involved in the growth of spores and spore-bearing structures. The whole process of reproduction, of whatever type, takes place in an orderly and more or less constant sequence of stages. It involves cell division and often coalescence of cells or adhesion of cell walls, cell enlargement, the alteration of cell shape, or of cell-wall thickness and structure, and the accumulation and differentiation of cell contents. Often the cells making up a mature reproductive body are entirely different in appearance from those of the vegetative mycelium or from those of immature reproductive structures. Those comprising a single fruit-body may be of several distinct types. Little is known of the mechanism of the actual shaping of the cells of spores and sporophores, but a few outstanding studies of the sequence of development and a few attempts to correlate visible development with the chemical and physical changes in the cell wall have been made.

SPORES

Mature spores show a wide range of size, shape, cell numbers, and pigmentation and many bear complex appendages or have elaborately sculptured walls. Spore characters are, however, remarkably uniform for any particular species. The majority of spores are colourless, single-celled, more or less globose or ovoid, thin-walled structures, and all of them pass through such a stage in the early phases of development. Fig. 1 shows the development of complex spores of various species from a single cell. We know little of the forces operating to mould the shape of spores. It is clear that if the spore is not subjected to unequal lateral pressure it is likely to remain more or less globose, provided that the spore wall develops and hardens uniformly. If, however, the developing spore is subjected to pressure from

7

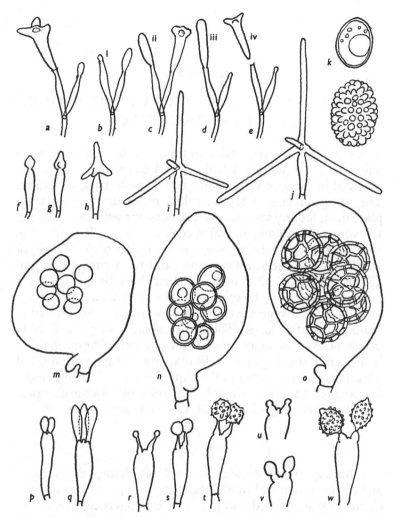

Fig. 1. Development of complex spores. *a–e*, *Heliscus aquaticus*, conidiophores with two phialides, drawn at intervals to show development of spores from each phialide alternately; i–iv show development of one such spore from small oval bud to the mature clove-shaped conidium; *f–j*, *Lemonniera aquatica*, single phialide drawn at intervals to show development of single conidium from more or less globose bud to complex four-armed spore (*a–j*, after Ingold, 1942). *k, l*, *Genea spherica*; *k*, immature ascospore with smooth wall; *l*, mature ascospore with hemispherical warts; *m–o*, *Hydnobolites cerebriformis*, stages in development of ascus and ascospores; *m* has immature spores with smooth thin walls; *n* has older spores, larger, with smooth thick walls; *o* has mature spores with thick, reticulately sculptured walls; *p, q,* *Hysterangium nephriticum*; *p*, young basidium with two young ovoid spores; *q*, mature basidium with three elongated spores with claw-like bases (number of spores on the basidium varies in this species); *r–t, Hydnangium carneum* var. *xanthosporum*; *r*, young basidium with small, smooth-walled, spherical spores; *s*, older basidium, spores still smooth-walled but larger; *t*, mature basidium, spores with spiny outgrowths from wall; *u–w*, *Hymenogaster tener*; *u*, apex of young basidium with small, smooth-walled spherical spores; *v*, older basidium, spores larger, ovoid, but still thin-walled; *w*, mature basidium, spores citriform, covered with blunt warts except at apex and base (*k–w*, after Hawker, 1954). ×375.

surrounding cells and if the wall again develops uniformly it is likely, on hardening, to conform to the shape of the available space between neighbouring units. Again, if part of the wall hardens and solidifies while cytoplasm is still flowing in from the supporting hypha, the parts which are last to harden are likely to expand, as when a plastic balloon is blown up while protruding from a solid constriction. The final shape of such a spore will depend upon the location of those parts of the wall which harden at different stages in spore development. The pattern of the deposition of wall-forming molecules is obviously of importance. Cylindrical spores may be formed as a result of continuous elongation after secondary deposition of wall material begins, if the elements are laid down in such a way that intussusception of new elements can lead to expansion in one direction only.

By measuring the length and breadth of developing basidia and basidiospores of species of *Clavaria* and other Basidiomycetes, Corner (1947) showed that these all begin as globose structures and develop in a manner characteristic of each species. He concluded that, where these remain spherical, the surface increases in area evenly by uniform intussusception of new material. Where expansion continues at the apex it may be concluded that the polar cap remains elastic after the rest of the cell wall has become rigid. When the spore becomes cylindrical, the rate of advancing fixation of the wall towards the apex must balance that of enlargement of the apex, so that the latter never attains a diameter greater than that of the rest of the cell. If, however, the rate of fixation of the wall is more rapid than that of apical enlargement, a cell tapering towards the apex results, while conversely the cell tapers to the base. It has been further suggested that rigidity of the wall is due to chitinization of the original hemicellulose or cellulose structure.

When a developing cell is subjected to unequal pressure typical shapes are assumed. Thus among Ascomycetes, forms with globose asci tend to have globose, or nearly globose, spores, as with some yeasts, with certain members of the Plectascales, such as *Elaphomyces* spp., or with species of *Tuber*. Forms with clavate asci often have elongated spores (e.g. *Claviceps* and *Geoglossum*), while in the cylindrical, close-packed asci of the Pezizales, in which the spores are usually arranged uniseriately,

9

the latter are commonly ellipsoidal and arranged at an angle to the long axis of the ascus, thus occupying the available space economically. There are, of course, many exceptions, such as the 'hat'-shaped spores of species of *Endomyces* and related genera, and the peculiar spores of many species of *Eurotium*. Both *Endomyces* and *Eurotium* have globose asci, and here one must suppose that a differential rate in wall fixation also plays a part as suggested by Corner (*loc. cit.*) for the Basidiomycetes. Ingold (1954) commented on the frequent occurrence of ascospores with a blunt apical end and a tapering base and suggested a possible biological advantage of this shape in facilitating discharge from the ascus.

Among the Hymenomycetes and Gasteromycetes, where the hymenium is extensive, the basidiospores are often more or less elongated with the long axis parallel to, or at a slight angle from, the axis of the basidium, i.e. at right angles to the plane of the hymenium. Here again one may suppose that the pressure of surrounding spores, which are often very numerous, has a formative effect. It may be remarked that, among hypogeous Gasteromycetes, those forms which shed their spores into relatively large cavities at maturity, e.g. *Hydnangium*, *Arcangeliella*, *Stephanospora*, have more or less spherical spores, while those which, like *Hymenogaster* and *Hysterangium*, retain the spores still attached to the basidia by the sterigmata for some time after maturity, tend to produce elongated spindle-shaped spores (see fig. 1).

The development of pigment and wall-sculpturing in spores must depend upon the accumulation of food materials within the spore, or on the availability of reserves in the matrix surrounding the spore, as with the cell sap of sporangia or asci. In *Bulgaria inquinans* when the nuclear division within the ascus is not simultaneous, the spores developing first may be pigmented, those developing later may be colourless, but still viable, or may abort. This is almost certainly due to competition for the available foodstuffs within the ascus. Similarly, it has been pointed out (Hawker, 1954, 1955) that nuclear divisions in the asci of truffles (*Tuber* spp.) are almost certainly not simultaneous, with the result that some of the nuclei abort; the asci thus contain irregular numbers of spores. Abortive spores of these fungi may be seen in young asci and the degree of sculpturing of the walls of the mature spores of some species may vary within the same ascus, presumably as a result of unequal utilization of the food materials in the cell sap. It will be noted in later chapters that the size and shape of the conidia of some fungi may be altered by differences in nutrition or some other factors (pp. 29, 32, 43, 60).

Thus it may be concluded that the final form of a spore depends upon its genetical make-up, upon the presence or absence of unequal pressure during development, upon the degree of uniformity in rate of cell-wall fixation over the surface area and on the available food supply and perhaps on other environmental factors such as aeration or humidity. A study of spore-wall deposition and final structure by the aid of modern methods is desirable. Studies, such as those of Hughes (1953), of the exact method of conidium formation in the Hyphomycetes, although undertaken in order to elucidate taxonomy, may provide the basis of physiological interpretation of spore formation.

SPORANGIOPHORES AND SPORANGIA OF PHYCOMYCES

The most detailed study of the stages in growth and the relation between these and wall structure has been made with the sporangium and sporangiophore of *Phycomyces blakesleeanus* (Castle, 1936, 1937, 1938, 1942, 1953; Errera, 1884; Graser, 1919; Heyn, 1936, 1939; Oort, 1931; Oort and Roelofsen, 1932; Pop, 1938). This is a comparatively simple object for study but the problem of its interpretation is still far from being solved. Development of the sporangiophore and sporangium of *Phycomyces* may be divided into four main stages (fig. 2). The first stage, which occupies about 1 hour under optimum conditions, is one of elongation of the growing zone just behind the tip of the sporangiophore. The second phase begins with the swelling of the tip of the sporangiophore to produce the spherical sporangium. Soon after the sporangium is initiated, growth in length of the sporangiophore ceases temporarily. When the sporangium has attained maximum diameter, growth more or less ceases for 2 to 3 hours. This is the third stage. Growth in length of the sporangiophore is then resumed, at first in a dextral (stage IV*a*) and then in a sinistral (stage IV*b*) spiral direction. The interpretation of this spiral growth of the sporangiophore has been the subject of much discussion and no satisfactory conclusions have yet been reached. Some observers, such as

Frey-Wyssling (1952), have attempted to interpret it as the result of spiral flow of protoplasm within the cell, leading to a spiral deposition of long chitin molecules. Others (Preston, 1952; Roelofsen, 1950, 1950a, 1951) consider it to be the result of the existing orientation of the primary wall structure, although

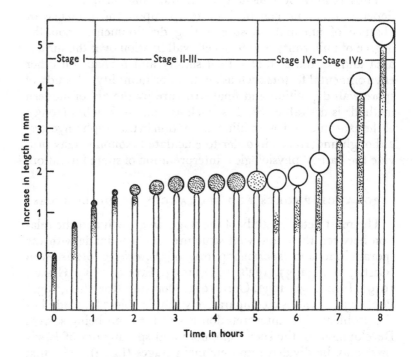

Fig. 2. Growth of sporangia and sporangiophores of *Phycomyces blakesleeanus*. Stippled areas indicate regions of active growth. Stage I = period of elongation of sporangiophore. Stage II = period of little or no elongation of sporangiophore, but of increase in size of sporangium. Stage III = period of little or no growth. Stages IVa and IVb = periods of resumed growth in length of sporangiophore in dextral and sinistral directions respectively (after Castle, 1942).

those authors differ in their theories of the means by which secondary deposition of wall elements is brought about. The one point about which all observers agree is that the spiral growth is correlated with a spiral arrangement of molecules in the wall as shown by electron microscopy and by other techniques. The subject has been discussed in detail by Castle (1953).

Since interpretation of such a relatively simple example as *Phycomyces* has proved to be so difficult it is not surprising that very little progress has been made in the determination of the formative factors controlling the development of larger and more complex structures such as fruit-bodies and sclerotia. Mycological literature abounds with detailed accounts of the development of these. Corner (1929, 1929a, 1932, 1932a, 1934, 1948, 1950) has examined the exact derivation of different parts of a number of ascomycetous and basidiomycetous fruit-bodies. In the comparatively simple apothecia of the Discomycetes (Corner, 1929, 1929a), he showed that marginal growth of the fruit-bodies takes place from fascicles of hyphae situated between the lower palisade layer, or 'cortex', and the upper palisade layer, the hymenium. He states that 'every hypha in the growing part arises as a lateral 2–3 μ wide, the growth-rate of which rises to a maximum, then falls, the fall being accompanied by branching in the proximal part and maturing of the distal part. The distal parts are transformed on the outside of the growing point into cortical hyphae, on the inside into paraphyses; the proximal parts form the medullary hyphae.' The growth of the margin is thus sympodial. Corner points out that the growing apices are fed from the medulla and that their elongation may cease as a result of reaching the limit of the nutritional range. A simple formative influence is thus suggested by the exact study of morphology in this type of fruit-body. Corner further shows that in addition to this marginal growth of the apothecial cup, the axis or stem, more particularly in stipitate species, is defined by an upward growth and that the final length of the stem is determined indirectly by a phase of cell enlargement following cell formation. In the interpretation of the effect of environment on fruit-body form it is clear, therefore, that attention must be paid to the effects on both cell formation and cell extension.

Corner has carried his technique of hyphal analysis still further in studying the more complex fruit-bodies of the higher Basidiomycetes. He introduces the terms monomitic, dimitic, trimitic, etc., for fruit-bodies with one, two, three or more systems of hyphae. Some fruit-bodies, such as those of *Poly-stictus xanthopus*, have as many as four systems of hyphae. In this

13

example these comprise *skeletal hyphae*, which are unbranched, thick-walled, aseptate and orientated in a longitudinal direction; *generative hyphae*, which are branched, thin-walled, septate, with clamp connexions, and make up the main bulk of the fruit-body; *binding hyphae*, which are branched, aseptate, thick-walled and interwoven between the other hyphae; and *mediate hyphae*, which are sparingly branched, aseptate, thick-walled and arranged longitudinally. The final shape and position of the fruit-body is determined by the reaction of the growing hyphal apices to external stimuli such as light and gravity. In *Fomes levigatus* the tips of the skeletal hyphae alone are the geotropically sensitive elements. In the spiny fruit-body of *Asterodon*, Corner (1948) showed that growth of the hyphae of the young fruit-body is at first outwards in all directions to form a loosely interwoven mass compacted by the arms of the stellate hyphae characteristic of this genus. The uniform outgrowth later slows down and the developing stellate hyphae are smaller and the generative hyphae more freely branched. Downward growth then stops and the end cells of the generative hyphae, together with sympodially intercalated new elements, form the incipient hymenium. In certain areas of this all hyphal tips become positively geotropic and a localized downward growth results, leading to the formation of spines. Pore growth in certain polypores is the result of a comparable localization of downward growth. Such studies are an essential preliminary to the analysis of factors determining growth form.

Bonner *et al.* (1956) studied growth of the common mushroom. Orientation of the component hyphae is followed, at an early stage in development, by elongation through cell expansion with little increase in cell number. Dry weight continues to increase, indicating the continued transport of food materials from the soil.

The factors controlling the proportions of complex fruit-bodies are not yet understood but it is clear that these are restricted to a certain range of relative sizes. Thus Corner (1934) showed a correlation between length of stipe and elongation of cells in the mature stems of certain agarics, between width of stipe apex and number of primary gills and between general diameter of stipe and number of secondary or branch gills. Ingold (1946) pointed out that form and size are related in agarics. The ratio of pileus diameter to stipe length increases

14

with increase in size of pileus and it is suggested that this is a result of mechanical laws.

Little is known of the method of development of stromata, but the formation of sclerotia has been studied for a number of species. Townsend and Willetts (1954) showed that sclerotia of different species are formed in very different ways. Thus those of *Rhizoctonia solani* develop by increased branching and septation of ordinary hyphae in a hyphal aggregate and by the inflation of the individual cells. Those of *Botrytis cinerea, B. allii* and *Sclerotium cepivorum* develop by the repeated dichotomous branching of a single hyphal tip, followed by increased septation and fusion of the branches, and finally by a differentiation of the outer cells of the resulting mass to give a protective layer. Those of *Sclerotium rolfsii* and *Sclerotinia gladioli* develop by the coalescence of branches put out from mycelial strands. With such diverse methods of development one would expect that the factors inducing the initiation, and controlling the further development, of the sclerotia of various species would be far from uniform. As will be seen in the next chapter this expectation is realized.

It is clear from these few examples that a detailed analysis of the nature and mode of growth of the various hyphal elements making up a complex structure is a valuable and essential preliminary to a study of the formative effects of environmental factors.

CHAPTER 3

THE PHYSIOLOGY OF VEGETATIVE
REPRODUCTION

THE simplest form of reproduction in fungi is the multiplication
and separation of individual cells to form the so-called 'sprout
mycelium'. This is best known among the true yeasts where it
is the dominant growth-form. Even here, however, certain
conditions will induce the formation of a 'pseudomycelium'
consisting of more elongated cells of irregular size, loosely joined
together in a chain. The pseudomycelial phase is considered by
Scherr and Weaver (1953) to be half-way towards the formation of a true mycelium. These authors further suggest that the
elongation of the cells of a sprout mycelium is a step towards the
formation of pseudomycelium. Fungi, such as species of *Endomyces*, which are obviously related to the yeasts, produce normal
filamentous hyphae as the dominant growth-form but break
down more or less readily into pseudomycelium, oidia and
typical sprout mycelia with ageing of a culture or under certain
environmental conditions. Many typically filamentous fungi,
such as certain species of *Mucor* or *Aspergillus*, which are not
closely related to the yeasts, also tend to produce oidia under
particular conditions, notably in liquid culture with a high
concentration of glucose or other sugar.

OIDIA

The exact circumstances controlling the production of
pseudomycelia and oidia are known in very few examples,
although much information of an empirical kind has been
amassed. The importance of more exact knowledge is obvious
when it is remembered that Lodder (1934) distinguishes between groups of asporogenous yeasts by means of their ability
to produce pseudomycelia. It is generally assumed that the
formation of oidia and sprout mycelia takes place at higher
nutritional levels than those leading to the formation of pseudomycelia. Thus in the well-known dimorphism shown by certain

16

animal (including human) pathogens, the organism is found only in the yeast-like or oidial form within the host, but develops hyphae in artificial culture (Ainsworth, 1952). In contrast, certain plant parasites, such as the smuts and species of *Taphrina*, which are normally mycelial in habit in the host plant, tend to produce only yeast-like masses of oidia in artificial culture. *Candida albicans* produces only yeast-like cells when the source of carbon is glucose, but when this is replaced by soluble starch, growth is slower and tends to become pseudomycelial in form (Ingram, 1955). It is well known that high concentrations of sugar lead to the production of oidia by many fungi. On the other hand, young cultures of *Endomycopsis capsularis* on malt agar are filamentous in form but break down into oidia as the culture ages and, presumably, the medium becomes partially depleted of nutrients.

High temperatures also favour the production of oidia in some species, as with the human pathogen, *Blastomyces dermatitidis* (Levine and Ordal, 1946), which produces only yeast-like cells at 37°C. but is entirely filamentous at 25°C. This effect of temperature is independent of nutritional factors.

Ingram (*loc. cit.*) also suggests that conditions of partial anaerobiosis, as in a liquid culture, may favour the production of pseudomycelia rather than of sprout mycelia in certain yeasts. This is certainly not universally applicable. It is well known since the early works of Pasteur (1876) that many species of *Mucor* and *Aspergillus* form oidia more readily in liquid than in solid media.

In view of the contradictory nature of the available information on the factors leading to oidial formation, studies such as that of Nickerson (1954) are to be welcomed. Nickerson points out that 'the fact that a particular morphogenetic development can be related to one component of the environment is insufficiently known'. In a number of examples, including parasitic forms (such as *Candida albicans*) and saprophytic species (such as *Mucor*), he puts forward schemes to show possible correlations between slight alterations in the metabolic pathway and changes in form. A filamentous mutant of *C. albicans* was shown to be maintained in this state by a block in a particular stage of metabolism, resulting in interference with an enzyme system normally controlling a step essential for the process of cell division. A similar block might result in the normal strain, possibly through exhaustion of the nutrient or by the action of specific inhibitors,

17

and would prevent the normal formation of sprout cells. An extension of study on these lines should lead to important results and the ultimate interpretation of the metabolic processes determining cell multiplication.

CHLAMYDOSPORES

The factors controlling the production of chlamydospores are even less well known than those determining the formation of oidia. *Candida albicans* is said to form chlamydospores only when the concentration of glucose in the medium is low (Nickerson and Mankowski, 1953; Ingram, 1955). A synthetic medium containing a polysaccharide (such as soluble starch purified to remove reducing sugar), as source of carbon, and ammonium sulphate, as nitrogen source, favours chlamydospore production. The addition of either glucose or cysteine to this medium prevents the formation of either filamentous hyphae or chlamydospores. The factors favouring chlamydospore formation are here the opposite of those leading to oidial production. With *Mucor racemosus*, however, it is common laboratory practice to induce the formation of both oidia and chlamydospores by the use of media of high sugar concentration. Oidia of this fungus are more frequent in liquid media, but chlamydospores are found in both liquid and solid media, provided the initial sugar concentration is high. They usually develop at a later stage than do the oidia. The chlamydospores of species of *Fusarium* are generally found in old cultures where the medium is almost certainly depleted of essential nutrients, but no exact data on this are available.

It is thus indicated that in these various species chlamydospores develop on a well-nourished mycelium only in the absence of hexose sugar, or when this has become exhausted by growth of the fungus. The chlamydospores of *Sepedonium chrysogenum*, which is parasitic on a number of basidiomycetous fruit-bodies, develop in enormous quantities on the decaying sporophores. A strain isolated by the writer from the hypogeous Gasteromycete *Melanogaster variegatus* produces numerous chlamydospores immediately after inoculation on to malt agar or other rich media. The so-called chlamydospores of the smuts and species of *Taphrina* develop in certain parts of the host plant at a particular stage in the process of parasitism, but whether this is

18

controlled by exhaustion of nutrients, or by entry of the mycelium into areas particularly rich in foodstuffs, such as ovaries or anthers, is not determined. These spores, however, despite their mode of formation, are not typical chlamydospores since, on germination, they give rise to basidia (promycelia) or asci respectively. Since typical chlamydospores have all the characteristics of resting bodies, such as thick walls, dense, oily contents, longevity and limited powers of germination, it is likely that exhaustion of food supply normally plays a part in their initiation. Venkat Ram (1952) reports that the addition of strains of certain bacteria to the soil increases the number of chlamydospores produced by *Fusarium solani*. He attributes this to the effect of the production of antibiotic substances by the bacteria. It would be of interest to know if other factors unfavourable to mycelial growth also hasten the development of chlamydospores.

The gemmae, or chlamydospores, of the water moulds (Saprolegniales) are produced when a well-nourished mycelium is transferred to a weak solution of sodium chloride or tricalcium phosphate, etc., or when the mycelium is left in an originally good nutrient medium until the nutrient is completely exhausted (Klebs, 1898; Coker, 1923).

SCLEROTIA

While the scanty data available suggest that chlamydospores form only on a well-nourished mycelium, the latter is obviously still more essential for the formation of the much more complex sclerotia. Sclerotia of different species differ greatly in size, shape and density of the hyphal components. Moreover, these characters may vary when a particular fungus is subjected to a variety of conditions.

Several authors have reported the effect of temperature on production of sclerotia (Paul, 1929, on *Botrytis cinerea*; Rogers, 1939, and Ezekiel *et al.* 1934, on *Phymatotrichum omnivorum*; Townsend, 1952, on *Botrytis cinerea*, *B. allii*, *Sclerotinia gladioli*, *Sclerotium rolfsii* and *Rhizoctonia solani*), and the general conclusion is that sclerotia form over the same range of temperature as that permitting mycelial growth, but that the time which elapses before visible initials are present and the rate at which these develop into mature sclerotia are greatly influenced by temperature. Moreover, the optimum temperatures for growth of sclerotia and mycelial production are not always identical. If the initiation and maturation of sclerotia are controlled by the synthesis of 'sclerotium-producing' and 'maturation' factors by the mycelium, then the effect of temperature on sclerotium production might be due to its effect on the rate of synthesis of such factors. The temperature favouring these processes would not necessarily coincide with that favouring growth as a whole.

Little is known of the effect of light on the initiation and maturation of sclerotia. Townsend (*loc. cit.*) carried out a few experiments on the effect of light on *Botrytis cinerea, Sclerotium rolfsii* and *Rhizoctonia solani.* The method used was crude and the possibility of differences in temperature and humidity between cultures in the light and the dark was not excluded. Nevertheless, her results are of sufficient interest to be worth recording. Ordinary daylight had no effect on *Botrytis cinerea* when compared with continuous darkness, but continuous light stimulated mycelial growth and depressed sclerotium formation. This agreed with the results of Paul (1929) for *B. cinerea.* With *Sclerotium rolfsii* the reverse effect occurred and continuous light accelerated the production of sclerotia and depressed mycelial growth. Sclerotia of *Rhizoctonia solani* formed earlier in the dark but matured more rapidly in the light, although the final number was not influenced. These results suggest that the effect of light is by no means simple, and further experiments with a more exact light control are desirable.

The effects of soil moisture, relative humidity and aeration on the production of sclerotia have not been studied in detail and, in general, it is assumed that the requirements are similar to those of vegetative growth. The effect of H-ion concentration of the substrate has, however, received more attention. Townsend (*loc. cit.*) studied the effect of initial pH of the medium on the production of sclerotia by a number of fungi and concluded that sclerotia developed over the same range as that permitting mycelial growth but that development was late and slow at the extreme limits of this range.

The effect of nutrition on the initiation and maturation of sclerotia has been extensively studied, but little or no attempt has been made to interpret the results obtained, which remain, therefore, as a mass of empirical data.

20

The source of carbon favouring the production of sclerotia is usually, but not always, the same as that giving maximum mycelial growth. Sources of carbon which permit only a starvation type of growth do not allow the formation of sclerotia. Concentration of carbohydrate is also an important factor but the results available are conflicting and insufficient to allow generalization. Moreover, Curzi (1932, 1932 a) showed that the effect of concentration of glucose may differ for strains of the same fungus, *Sclerotium rolfsii*. A strain from potato produced large numbers of sclerotia of normal shape and size on media with high concentration of glucose, while a strain isolated from asters produced fewer, but larger and irregularly shaped sclerotia with increase in glucose concentration.

The effect of the source of nitrogen has also been studied by a number of investigators. Thus Higgins (1927) concluded that organic nitrogen is necessary for good growth and production of sclerotia by *S. rolfsii* which did not readily utilize nitrogen in the form of nitrates or ammonium salts. Townsend (*loc. cit.*), however, concluded that ammonium salts are a satisfactory source of nitrogen for this fungus, provided the medium does not become too acid and if adequate growth-substances are present. All the fungi studied by her could use ammonium nitrogen, nitrate nitrogen and some form of organic nitrogen for growth and sclerotium production, provided that other environmental factors were not limiting. Urea was an unsatisfactory source of nitrogen but asparagine, peptone, casein and egg albumin were all suitable. Pape (1937) also found urea to inhibit the production of sclerotia by *Sclerotinia trifoliorum*. He found that ammonium salts caused the formation of large sclerotia while glycocoll induced the production of numerous minute ones. Peiris (1947) states that *Botrytis cinerea* is unable to form sclerotia with ammonium nitrate as sole source of nitrogen, but does so readily with peptone, asparagine or potassium nitrate. Démétriadès (1953) showed that potassium nitrate or certain organic nitrogenous substances (glycine, asparagine, aspartic and glutamic acids, tyrosine, peptone and urea) favour the production of sclerotia by *Sclerotinia sclerotiorum*, while few sclerotia are formed with ammonium salts or leucine as source of nitrogen, and none are produced with cysteine, cystine, lysine, methionine, tryptophane or valine in the absence of other nitrogenous substances.

Peiris (1947) demonstrated the importance of the carbon-nitrogen ratio in the formation of sclerotia. He showed that, with *Botrytis cinerea*, there is an increase in the number of sclerotia with increase in glucose concentration, provided that sufficient nitrogen (in the form of peptone) is present, but that increase in nitrogen inhibits the production of sclerotia while increasing that of conidia. Townsend (*loc. cit.*) repeated this work and extended it to a study of other sources of nitrogen and of other fungi. Provided that there was an adequate source of nitrogen in a suitable form, sclerotial production always increased with increase in glucose concentration. Garrett (1949) attributed the depressing effect of excess nitrogen on the production of sclerotia by *Helicobasidium purpureum* to the resulting increase in mycelial growth, which in turn reduces the level of carbohydrate in the medium below that necessary for maturation of viable sclerotia. In a personal communication he reports that numerous sclerotial initials developed on media with a high concentration of nitrogen, but that these failed to mature. Allington (1936) claimed that the number of sclerotia produced by *Rhizoctonia solani* increased with decrease in carbon, and decreased with decrease in nitrogen content. His results are therefore directly opposed to those of Peiris, Garrett and Townsend and it is difficult to see how they could be reconciled. A difference in the strain used, or in other environmental factors, may have influenced the apparent effect of the concentration of carbon and nitrogen. Work now in progress, however, confirms Townsend's results in this respect.

Earlier writers have claimed striking effects on sclerotial production from the use of natural extracts and some have attributed this to the presence of unknown stimulating substances (e.g. Nakata, 1929). Robbins and Kavanagh (1938) claimed that thiamin (vitamin B_1) induced the formation of numerous sclerotia by *Sclerotium rolfsii* and that the pyrimidine and thiazole components of this vitamin together had a similar effect. Pyrimidine alone was almost as good, but no sclerotia formed on thiazole alone in the absence of pyrimidine. Townsend (*loc. cit.*) confirmed the effect of thiamin on this fungus. Lilly and Barnett (1948) studied the interaction of growth rates, vitamin deficiencies and sclerotial formation in a number of species of Sclerotiniaceae and concluded that the effect on the latter was inversely proportional to the ability of a particular

22

fungus to synthesize the necessary vitamins. The known vitamins do not, therefore, have a specific effect on sclerotial production, but where a fungus either cannot synthesize a particular vitamin or does so only slowly, an external supply is essential for both growth and sclerotial production. Occasionally, however, a fungus synthesizes enough for good mycelial growth but needs an additional external supply before sclerotia will develop. The needs of these complex bodies are obviously greater than those of vegetative hyphae.

While known growth-substances influence the production of sclerotia in a manner parallel to their effect on vegetative growth, the possibility is by no means excluded that specific substances do exercise control of the process of initiation and maturation of these bodies. Moreover, *the processes of initiation and maturation are not favoured by the same set of environmental conditions.* In Townsend's experiments on the carbon-nitrogen ratio some sets of conditions gave numerous sclerotial initials which did not mature and others induced the formation of a few large and mature sclerotia. The reason for this difference was not investigated but it is not impossible that maturing sclerotia inhibit the development of initials, either by competition for food, as with the formation of rhizomorphs by *Armillaria mellea* (Garrett, 1953), or by the actual production of inhibiting substances.

On the whole it may be said that sclerotia develop under conditions leading to optimal growth but that their food requirements, as might be expected from a consideration of their bulk and their rich reserve food contents, are greater than those of the mycelium.[1] The food supply must not be exhausted too soon if the sclerotia are to mature. One difficulty in the interpretation of results on sclerotia is lack of exact knowledge of the significance of these structures. Some of them ultimately bear the fructifications of the fungus and their formation is thus a first step to fruiting, but others have never been known to bear spores of any type and are solely organs of vegetative reproduction and survival. While the term 'sclerotium' covers such diverse bodies, no generalization can be made relating to the physiology of sclerotial formation as a whole, but only relating to members of a particular group such as species of *Sclerotinia* and *Botrytis* or of *Rhizoctonia*.

[1] Ergle and Blank (1947) showed that total carbohydrates of mycelium and sclerotia of *Phymatotrichum omnivorum* represented 23 and 46% respectively of the dry weights.

CHAPTER 4

THE EFFECT OF ENVIRONMENT ON SPORULATION

THE conditions which permit spore formation are almost always of narrower range than those permitting mycelial growth. Thus, in general, spores are produced over narrower ranges of temperature, humidity, water content and hydrogen-ion concentration of the substrate than those within which vegetative growth is possible. Many fungi which can grow vegetatively in the dark are unable to produce their spores in the absence of light. Conditions permitting the development of the 'perfect' spore stage usually differ far more from those sufficient for vegetative growth than do those favouring asexual reproduction. The effect of nutritional factors is so important and varies so much with different species and different types of spore that a consideration of nutritional effects will be deferred to the next chapter.

TEMPERATURE

It is well known that mycelial growth of a fungal colony is most rapid at a particular temperature which may be called the optimum and which is usually in the region of 18–25°C., although some fungi may have an optimum temperature below or above this. There is also a minimum temperature, below which growth ceases, and a maximum above which it also ceases. The maximum temperature is often nearer to the optimum than is the minimum. These three temperatures, minimum, optimum and maximum, which are called the cardinal points, vary not only from one species to another but among races or strains of the same species. Moreover, they may also vary for the same strain under different environmental or nutritional conditions.

For a particular strain under a particular set of conditions, however, it is usual for the range of temperature permitting sporulation to be narrower than that allowing mycelial growth, that is, growth may take place at a temperature sufficiently low to inhibit sporulation, and at the other end of the range mycelial growth may still continue (although the hyphae are often of an abnormal form) at a temperature above that at which spore formation ceases. Table 1 gives examples of the ranges of temperature permitting vegetative and reproductive growth of a number of fungi. The optimum temperature may also be different for growth and sporulation of the same fungus (Table 1). The temperature optimal for sporulation may be different on different media. *Melanospora destruens*[1] (Hawker, 1947) has a lower optimum for growth and a higher one for production of perithecia on a medium containing sucrose as the sole source of carbon, than on a similar glucose medium. It is suggested that sucrose, which is respired more rapidly than glucose, is exhausted more quickly than glucose at high temperatures, so that the resulting low concentration of sugar favours fruiting but limits growth.

The effects of small differences in temperature are often great enough to cause zones of intense sporulation alternating with zones of sparse or no spore production in colonies growing under conditions of fluctuating temperature, as with *Fusarium fructigenum* (Brown, 1925) or *Ascochyta rabiei* (Hafiz, 1951).

With those fungi that produce more than one type of spore the optimum temperature is seldom the same for the production of all types; the corresponding maximum and minimum temperatures may also differ. Thus *Sporodinia* (*Syzygites*) *grandis* (Baker, 1931) produces zygospores at higher temperatures than those that favour the development of sporangia. On the other hand sporangia of *Mucor sexualis* and of some other members of the Mucorales are able to develop at temperatures that are too low for production of zygospores. That the effect of temperature on spore production in the Mucorales is extremely complex is shown by the work of Barnett and Lilly (1950), who report that *Choanephora cucurbitarum* produces only conidia in the natural habitat, but produces both conidia and multispored sporangia in culture at relatively high temperatures. Thus at 25° C. conidia are more numerous than sporangia, but at 30° C. the reverse is true and at 31° C. only sporangia are formed.

Similarly, among Ascomycetes, the temperature requirements for the production of conidia may differ from those for the production of ascospores. The cleistocarps of *Eurotium herbariorum*

[1] Since renamed *Sordaria destruens* (Hawker, 1951).

TABLE 1. *Temperature ranges for growth and sporulation in certain fungi.* (Temperature given in °C.)

Fungus	Growth			Sporulation			Type of sporulation
	Min.	Opt.	Max.	Min.	Opt.	Max.	
Eurotium repens (Klebs, 1900)	7–8	—	37–38	8–9	—	35–36	Conidia
				—	—	33–34	Cleistocarps
Sporodinia grandis (Klebs, 1900)	1–2	—	31–32	5–6	—	29–30	Sporangia
				5–6	—	27–28	Zygospores
Saprolegnia mixta (Klebs, 1900)	0–1	—	36–37	1–2	—	32–33	Zoosporangia
				1–2	—	26–27	Oospores
Pilobolus microsporus (Gräntz, 1898)	2–4	—	33–34	10–12	—	28–30	Sporangia
Mucor racemosus (Klebs, 1900)	4–5	—	32–33	6–7	—	30–31	Sporangia
M. sexualis (Hawker, unpublished)	1	—	—	5	—	—	Sporangia
				10	—	—	Zygospores
Penicillium sp. (Wiesner, 1873)	2·5	—	43	3	—	40	Conidia
Saccharomyces cerevisiae (Hansen, 1883)	4	—	38	9	—	37·5	Asci
Gnomonia vulgaris (Henriksson and Morgan-Jones, 1951)	5+	17	30	10+	15	19–25	Perithecia
G. intermedia (Henriksson and Morgan-Jones, 1951)	5+	19	35	5	15–19	19	Perithecia

A dash (—) means no data.

26

are produced most readily on a suitable medium at temperatures of 25–30°C. while the conidial (*Aspergillus*) stage occurs also at lower temperatures. Barnett and Lilly (1947 *a*) showed that conidia of *Ceratostomella fimbriata* are formed at 18°C., but that no perithecia develop at this temperature. Numerous perithecia containing viable ascospores are produced at 25°C.

Thus there is no general rule as to the differential effect of temperature on asexual and sexual spore stages, since some fungi produce their sexual spores at relatively low temperatures and others only at higher temperatures than those permitting asexual reproduction. It is highly desirable that the mechanism of the effect of temperature should be studied in a number of further examples. One might expect that the effect of increased temperature would be to accelerate synthetic processes and thus to favour the production of the more complex sexual spores and fruit-bodies, which may be the case with *Eurotium* and *Ceratostomella* as cited above. It is equally probable that high temperatures might so accelerate vegetative growth and asexual reproduction that the available food material would be used up before the synthesis of complex growth-regulating substances had reached the point at which sexual reproduction could begin. This seems to be a likely explanation of the behaviour of *Sporodinia grandis*, in which the number of zygospores formed at relatively high temperatures increases markedly with an increase in the concentration of the medium (Baker, 1931). The final result is likely to be the result of both these effects in combination. The effect of temperature on various metabolic processes may well be so delicately balanced that small changes in other environmental factors actually reverse the temperature effect.

Temperature may sometimes act indirectly, as with *Phycomyces blakesleeanus*, in which Robbins and Schmidt (1945) demonstrated that temperature influences zygospore formation through alteration in the hydrogen-ion concentration of the substratum. There exists here a wide field for further investigation.

Not only does the effect of temperature vary with different types of sporulation of the same fungus, but it may also differ with particular stages in the production of the same spore or fruit-body. Thus Hirsch (1954) showed that if cultures of

27

Neurospora crassa are first incubated at 25°C. and then transferred to 35°C. after conidiation[1] numerous protoperithecia are formed, but none of these develop into mature perithecia. If colonies are incubated continuously at 35°C. no protoperithecia are produced. Colonies grown at 35°C. and transferred to 25°C. after conidiation and colonies kept continuously at 25°C. produce mature perithecia, while those kept continuously at 30°C. produce protoperithecia which do not develop further (see Table 2). Hirsch suggests that temperature influences at least two stages in development, that is the initiation of protoperithecia and some stage in their further development after conidiation which might be fertilization itself. No indication was obtained of the mechanism of this effect. Extracts of cultures bearing protoperithecia or perithecia had no obvious stimulating effect on cultures grown at 35°C., while extracts of the latter did not inhibit fructification in cultures at 25°C.

TABLE 2. *The influence of temperature on formation of protoperithecia and mature perithecia by* Neurospora crassa. (Data from Hirsch, 1954.)

Temperature of incubation before conidiation (°C.)	Temperature of incubation after conidiation (°C.)	Protoperithecia	Perithecia
25	25	+	+
30	30	+	−
35	35	−	−
25	35	+	−
35	25	+ (formed after transfer to 25°C.)	+

When cultures of *Mucor sexualis* and some other Mucorales on malt agar or glucose asparagine agar are transferred to temperatures below 7°C., zygospores develop only if the process of conjugation has reached the stage at which the gametangia are delimited by cross walls. No further conjugation takes place (Perkins, 1952; Roberts, 1954; and Hawker, unpublished data). Thus the inhibiting effect of low temperature acts only upon the early stages of zygospore production. In this example there is some evidence that the effect of low temperature is one of inhibition of some synthetic process, since young gametangia may continue to develop in the presence of old zygospores of the same or of a related species. Extracts prepared by various methods from old cultures are not effective, a fact suggesting that the

[1] In genetical work with this fungus crosses are made by adding conidia of one strain to young cultures of another. This is referred to as 'conidiation'.

28

hypothetical substance controlling further development is of a labile nature and is destroyed on exposure to air. The addition of small amounts of adenine or of some other purines induces a few, but not all, young gametangia to continue to develop. This effect is enhanced by the presence of thiamin (vitamin B_1). Biotin has no obvious effect. These substances cannot entirely replace the effect of living mature zygospores or of high temperature and it is obvious that the whole subject is exceedingly complex.

Some interesting results were obtained by Bowman (1946) with *Ustilago maydis*. He showed that fusions between sporidia of this fungus occur freely only over the narrow temperature range of 20–24° C. Lower temperatures greatly reduce conjugation and at temperatures above 24° C. conjugation is replaced by yeast-like budding of the sporidia. The two previous examples cited suggest that temperature requirements for nuclear fusions may be particularly exacting and this has been shown to be the case with *Ustilago*. Further attention might well be paid to the effect of temperature and of other environmental factors on actual nuclear fusion.

Temperature not only influences sporulation quantitatively and in some species controls the type of reproduction taking place, but it may also strongly influence the morphology of spores and spore-bearing structures. The size of the sporangia of *Choanephora cucurbitarum* (Barnett and Lilly, 1950) increases with increase in temperature, but the size of the conidia of *Peronospora parasitica* (Thung, 1926) is less ($23 \times 19 \cdot 5$ μ) at 20° C. than at 5° C. (27×23 μ). The relative size of ascus and ascospore of the yeasts, *Debaryomyces* spp. (Mrak and Bonar, 1938), changes with temperature so that at low temperatures the asci are not full and the warts on the ascospores are more readily seen than at high temperatures when the spores fill the ascus. Arens (1929) showed that both temperature and humidity influence the length of the sporangiophores of *Pseudoperonospora humuli*. The oogonia of *Achlya colorata* are papillate when formed at 15° C. and smooth-walled when formed at 20° C. and over (Reischer, 1949). Thus temperature may alter many characters usually regarded as diagnostic.

It is clear that temperature influences sporulation by its effects on a number of metabolic processes and by modifying the physical nature of the substrate.

Fungi in general grow in damp habitats and both the water content of the substrate and the humidity of the atmosphere must usually be fairly high to permit mycelial development. As a rule sporulation takes place only under wetter conditions than those sufficing for mycelial growth and particularly only with higher humidity. This, as might be expected, is most marked with the semi-aquatic members of the Phycomycetes and with those forms, such as Zygomycetes, that form thin-walled aerial sporangia, or many of the Hyphomycetes that produce relatively thin-walled conidia on aerial conidiophores.

Most of the chytrids and water moulds (Saprolegniales) occur submerged in water, but some of them grow in wet soil and some of the water moulds can be cultivated on agar media. With these, however, at least a film of free water is necessary to allow the production of zoosporangia, while some must be transferred to liquid culture before sporangia will develop. Wet conditions, or at least a very high humidity, are essential for the production of the sporangia of species of *Pythium* and *Phytophthora*. If slightly drier conditions follow, the 'sporangia' may germinate as conidia by means of a germ tube instead of producing zoospores. Waterhouse (1930) studied a number of species of *Phytophthora*. From a consideration of earlier work and from her own experimental results, she concluded that the presence of free water increases both the total number of sporangia (conidia) produced and the average number produced on a single sporangiophore. Even those species that produce sporangia freely on solid media do so more readily when submerged in water.

It is well known that the sporangiophores of *Phytophthora infestans* (potato blight) develop only during warm, damp nights and that the sporangiophores or conidiophores of the downy mildews similarly require a high relative humidity. Thus Yarwood (1943) showed that the conidia and conidiophores of *Peronospora destructor* (onion mildew) develop only over the range of 90–100% R.H., while those of *Bremia lactucae* (lettuce mildew) were shown by Ogilvie (1944) to require a relative humidity of 98–100%. Orth (1937) demonstrated that the development of sporangia of *Phytophthora infestans* is checked by quite brief fluctuations of 5% below the optimum humidity. The effect of fluctuating humidity is much more severe than that of large fluctuations in temperature.

Although many of the Zygomycetes and the majority of the conidium-producing Ascomycetes and Fungi Imperfecti grow in relatively dry habitats, compared with *Pythium* and *Phytophthora*, yet they all require fairly moist conditions for asexual sporulation to take place. Even members of the Erysiphaceae (powdery mildews), which are remarkable for their success under hot, dry conditions, require damp conditions for conidial formation, and Hashioka (1938) reported that *Sphaerotheca humuli* var. *fuliginea* produces abundant conidia over a range of R.H. of 93–96% but that the number produced is somewhat reduced at 100% R.H. Cheal and Dillon-Weston (1938) showed that a film of water is necessary to induce the budding-off of the conidia of the pear scab fungus, *Venturia* (*Fusicladium*) *pyrina*, and many such references may be found in the literature of plant pathology.

Where more than one type of asexual spore is produced by the same fungus the type of spore may be determined by the moisture conditions. For example, Barnett and Lilly (1955) showed that both temperature and relative humidity influence the relative number of conidial heads and multispored sporangia of *Choanephora cucurbitarum*. Thus at temperatures of 25° C. and over, a relative humidity of 100% favours the production of sporangia but at lower humidities conidia predominate.

While the development of the fleshy fruit-bodies of many Ascomycetes and Basidiomycetes is dependent on moist conditions, yet 'sexual' reproduction can sometimes take place under rather drier conditions than those needed for the production of asexual spores. Woody fruit-bodies such as those of *Ganoderma applanatum* (Hopp, 1938) may continue to develop in dry air and are in fact little influenced by changes in humidity.

It is probable that the sexual organs (oogonia and antheridia) of *Pythium* and *Phytophthora* can develop under slightly less wet conditions than those necessary for production of sporangia. Sexual reproduction of *Phytophthora cactorum* takes place freely on oatmeal agar in a closed petri dish, that is in air under conditions of high humidity, but sporangia are few unless the agar is flooded with water. Some other species behave similarly.

Sweet (1942) studied two species of *Magnusia* and showed that while the range of relative humidity permitting conidial

31

development is much less than that at which vegetative growth can take place, that permitting the development of cleistocarps is as great, or nearly as great, as the latter. The optimum (91·2% R.H.) was the same for all these processes. In contrast Klebs (1898) claimed that the optimum R.H. for zygospore production in *Sporodinia grandis* is greater than that for sporangial development, and that zygospores are produced in atmospheres near saturation point. This was confirmed by Robinson (1925), but Baker (1931) claimed that both spore forms occur over a range of humidities of 0–100%.

Atmospheric humidity has a profound effect also on the morphology of spores and spore-bearing structures. Walter (1924) showed that the rate of growth of sporangiophores of *Phycomyces nitens* is influenced by humidity. Changes in humidity are followed by changes in growth rate and this is reflected in the negative hydrotropic curvatures of sporangiophores, due to differential growth rates. Similarly, when fungi such as *Botrytis cinerea* or *Penicillium* sp. are grown in a saturated atmosphere, the conidiophores are long and branch indefinitely, producing few spores, while in a drier atmosphere they are short and bear numerous spores. The sporangiophores of *Mucor hiemalis* fail to develop in a dry atmosphere. The xeromorphic sporangiophores of *Phycomyces blakesleeanus*, however, develop normally in a dry atmosphere, provided the substrate is kept moist (Ingold, 1954*a*).

Gregory (1939) showed that the conidia of *Ramularia vallisumbrosae* formed under very wet conditions are shorter, broader and have fewer septa than those from drier situations. This is not a general response to humidity, since, with many species that produce septate spores, the length and the number of septa increase with increasing humidity.

Humidity may also influence the form of the fruit-bodies of Higher Fungi. The drying effect of a stream of air may cause abnormal tremelloid outgrowths from the pilei of certain agarics. Such abnormal fruit-bodies are occasionally found in nature and were produced artificially on the pilei· of *Coprinus ephemerus* by Keyworth (1942). Plunkett (1956) has studied the effect of the loss of water through transpiration on the development of fruit-bodies of *Polyporellus brumalis* and *Collybia velutipes*. The ratio of stipe length to diameter of cap of the former is greatest when loss of water is low. This species, however, is distinctly xeromorphic, since fruit-bodies are able to grow

32

to maturity and shed spores under very dry conditions. In contrast, both elongation of stipe and expansion of pileus of *Collybia* are inhibited when water loss rises above a rather low level. At levels permitting fruit-body development, the ratio of stipe length to pileus diameter in this fungus is not influenced by changes in humidity. With both these fungi, the effect of humidity may be modified by changes in other environmental factors.

The composition of the atmosphere may influence growth and sporulation not only through relative humidity, but by the relative concentrations of oxygen and carbon dioxide and the presence or absence of toxic gases, such as ammonia or certain volatile fungicides, e.g. the chlorinated nitrobenzenes. All these factors have a profound influence on the amount and form of fungal growth. It is common experience that poor aeration, such as that in a closed, tightly fitting petri dish or a plugged tube, reduces sporulation of many fungi. There are many references in the literature to the depressing effect of poor aeration on sporulation and to the more exacting requirements of sporulation as compared with mycelial growth, e.g. formation of pycnidia of certain members of the Sphaeropsidales (Leonian, 1924); of apothecia by species of *Ascobolus* (Green, 1930); of fruit-bodies by *Polyporus (Polyporellus) brumalis* (Bannerjee and Bakshi, 1944); but few attempts have been made to distinguish between the various factors which may operate in such conditions of poor aeration.

All fungi, with the possible exception of certain aquatic species, are obligate aerobes. Growth is inhibited by absence of oxygen, although some metabolic processes, such as alcoholic fermentation by yeasts, may continue. Nevertheless, vegetative growth often takes place in the presence of only minute traces of oxygen and is usually uninfluenced by a wide range of concentrations of this gas. Where the effect of oxygen concentration on sporulation has been examined it has been found that the requirements are higher than the minimum permitting vegetative growth. Thus Coons (1916) showed that the formation of pycnidia of *Plenodomus fuscomaculans* may be inhibited when there is still sufficient oxygen for growth of mycelium. Denny (1933) showed that, while colonies of *Neurospora sitophila*

33

continue to grow in the presence of only traces of oxygen, the production of perithecia is prevented at concentrations of less than 0·5%.

The accumulation of carbon dioxide has been shown to inhibit sporulation of a number of fungi in conditions of poor aeration, as with the failure of *Mucor mucedo* to produce sporangia (Lopriore, 1895) or of *Choanephora cucurbitarum* to produce either conidia or sporangia (Barnett and Lilly, 1955) under such conditions. Ascospore production by baker's yeast is also reduced by increased carbon-dioxide concentration (Bright *et al.* 1949).

Certain fungi, such as species of the hyphomycetous genera *Clathrosphaerina* and *Helicodendron*, which normally grow on decaying leaves in stagnant water, under conditions which must be very nearly or completely anaerobic, produce their spores only when the leaves are removed from water and exposed to moist air (van Beverwijk, 1951; Glen-Bott, 1955). These have been termed 'aero-aquatic' fungi. *Blastocladia pringsheimiana*, which also grows on various decaying organic matter in stagnant water, produces its thin-walled zoosporangia freely in aerobic conditions but produces the thick-walled resting sporangia only when the concentration of carbon dioxide is 99·5% or over (Emerson and Cantino, 1948). Such a concentration is probably often produced in the natural habitat by bacterial activity.

The inhibiting effect of poor aeration on sporulation of *Piricularia oryzae* is attributed by Henry and Andersen (1948) to accumulation of ammonia. Similarly the reduced sporulation of species of *Fusarium* in closed dishes, on media containing excess organic nitrogenous compounds, may also be due to the release of ammonia which is present in readily detectable amounts in such cultures (Brown, 1923).

Certain volatile fungicides, such as the proprietary forms of chlorinated nitrobenzenes, owe their success in controlling diseases such as the *Botrytis* disease of lettuces (Brown, 1935) and of other crop plants, to their inhibiting effect on sporulation, thus preventing the disease from spreading. Tetrachloronitrobenzene completely suppresses conidial formation in *B. cinerea* and reduces it in *Fusarium coeruleum* and *Trichoderma viride* (Reaver, 1954).

Aeration has a marked effect on the development of the fruitbodies of the Hymenomycetes. The effect of relative humidity and of a drying stream of air has already been referred to. Lambert

34

(1933) shows that if carbon dioxide is run under a bell jar inverted over growing sporophores of the cultivated mushroom, an accumulation of as little as 1% is injurious and 5% or more causes markedly abnormal growth, stunting and even death of the fruit-bodies. In contrast, excess oxygen gives heavier and more compact fruit-bodies. Mader (1943), however, claims that the injurious effect of poor aeration is due to volatile substances of unknown nature which accumulate in the air above the mushroom beds. Bannerjee and Bakshi (1944) claim that good aeration is essential for the normal development of the pileus of *Polyporellus brumalis*. More recently Plunkett (1956) has undertaken an accurate examination of the effects of aeration and other factors on the number and form of fruit-bodies of *Collybia velutipes* and *Polyporellus brumalis* in artificial culture. He also shows that aeration is essential for the normal expansion of the pileus and that if the pileus fails to expand the stipe continues to grow indefinitely. He shows that with adequate illumination, excess carbon dioxide is a possible cause of inhibition of pileus development.

It is clear that, in the absence of artificially introduced fungicidal gases, concentration of carbon dioxide is the most important factor controlling the effect of poor aeration on both the quantity and the morphology of spores and sporophores produced by fungi.

HYDROGEN-ION CONCENTRATION OF THE SUBSTRATE

Growth of fungi usually takes place over a fairly wide range of hydrogen-ion concentration and there is often no clearly marked optimum pH (Hawker, 1950). Some fungi are able to sporulate over the same range as that permitting mycelial growth, or over a slightly narrower range. Thus *Melanospora destruens* (Asthana and Hawker, 1936) produces perithecia freely over the same wide range of initial pH 4·8–7·6 as permits good mycelial growth, both growth and fruiting falling off rapidly beyond these limits. Others show a much more restricted range for sporulation than for growth. Growth of *Sordaria fimicola* (Lilly and Barnett, 1947) takes place slowly at an initial pH as low as 3·98, but perithecia are not formed when the initial pH is as low as 4·0, until nineteen days after inoculation, when internal changes in the medium have raised the pH to 6·84.

Lilly and Barnett concluded that a pH of at least 6·0 is essential for the initiation of perithecia. Henriksson and Morgan-Jones (1951) showed that the optimum initial pH values for the development of perithecia of two species of *Gnomonia* differ from those optimal for vegetative growth and from each other. While the effect of initial pH on sporulation is reported for a variety of fungi by different authors, few studies have given attention to the change in pH during growth. This actual pH, rather than the initial value, is the critical one and studies disregarding it are of limited value.

Even where fruit-bodies are formed over a comparatively wide range of pH they may not produce viable spores over the whole range. Lockwood (1937) showed that the percentage of cleistocarps of *Penicillium javanicum* and *Eurotium herbariorum*, and of perithecia of *Chaetomium globosum* containing fertile ascospores increases with increase in pH up to 7·0–8·0.

Plunkett (1953) and Aschan (1954) both found that fruit-bodies of *Collybia velutipes* are produced over a narrower range of pH than that permitting growth. The former shows that while fruit-body primordia fail to form in media with an initial pH of 4·5, yet with a slightly higher (5·0) initial pH the primordia that are formed will continue to develop after internal changes in the medium have caused the pH to fall to 4·2–4·4, indicating that the effect on the initiation of fruit-bodies differs from that on their further development. Aschan points out that, although, in her experiments, a pH of 6·0 was best for fruit-body production, yet under some circumstances a lower pH may induce earlier fruiting.

As with other environmental factors already discussed, pH may act differently on sexual and asexual stages of a particular fungus. Thus Lilly and Barnett (1951) noticed that *Eurotium rugulosum* produces numerous perithecia and a few conidia at an initial pH of 6·0–8·0, but at low initial pH values, perithecial formation is inhibited while conidia are produced freely.

Sørgel (1953) showed that the dominant reproductive phase of *Mycosphaerella pinodes* varied from pycnidia at pH 5·0 to pseudothecia at pH 7·0 and chlamydospores at pH 8·0. With decrease in pH below 5·0, reproduction is delayed, the size of the various reproductive organs is reduced and extensive aerial hyphae develop. An abrupt change in pH may result in a change in the nature of reproductive bodies already initiated, so that

36

pycnidial initials may develop into pseudothecia and vice versa. Such a change in the type of reproduction by the influence of an external factor is unusual, and it would be of interest if the work were extended to an examination of the mechanism concerned.

The apparent effect of other environmental factors, such as temperature or nutrition, is often found to be indirectly due to changes in pH during growth. This has already been pointed out for *Phycomyces blakesleeanus* (p. 27). The beneficial effect of calcium on sporulation, e.g. on perithecial production by *Chaetomium globosum* (Basu, 1951), may well be due to its effect in counteracting acidity. It is clear that, in any studies of sporulation, both initial pH and the changes that take place during growth must be taken into consideration in the interpretation of results.

VARIOUS TYPES OF INJURY

Toxins. Toxic substances used as fungicides often inhibit sporulation at a lower concentration than that needed to kill the fungus or even to stop growth. Since the prevention of sporulation and hence of dispersal of the causal fungus may be sufficient to control a plant disease, this is of practical importance and may permit the successful use of a fungicide at a level below that which would injure the host plant. The literature also contains many references to the inhibition of sporulation by the antibiotic substances produced by some other organism, under conditions which allow vegetative growth to continue.

Mechanical injury. Mechanical injury to the mycelium may lead to a local intensification of sporulation, as when cultures of *Fusarium fructigenum* are stroked with a needle, or when mycelium of *Pyronema confluens* (Robinson, 1926) is damaged by a knife or by the presence of crystals of toxic chemicals. Other fungi, e.g. *Melanospora destruens* (Asthana and Hawker, 1936), are not so stimulated. Shredding of cultures of *Alternaria solani* (Rands, 1917), or scraping of the mycelium followed by brief exposure to ultra-violet light (McCallan and Chan, 1944) induces increased production of conidia (see also p. 44).

The effect of moderate injury may be to release certain complex substances from the injured cells, which can be utilized in spore production. It is well known that sporulation often follows

37

the translocation of cell contents from the older parts of a colony (which subsequently die) to the younger hyphae.

Barriers to growth. Some fungi sporulate on reaching a mechanical barrier to growth, such as the edge of a petri dish. Asthana and Hawker (1936) showed that, while perithecia of *Melanospora destruens* are produced first at the edge of small petri dishes, they develop before the margin of the growing colony reaches the edge of dishes of 20 cm. diameter. These authors concluded that contact with the edge of the dish is not the main cause of perithecial formation but that age of the colony is more important. Robinson (1926), however, concluded that such a mechanical barrier definitely stimulates the development of the apothecia of *Pyronema confluens*. More recently (Buston, personal communication)[1] it has been suggested that the aggregation of perithecia of *Chaetomium globosum* at the edge of the petri dish is due to the prevention by this barrier of the diffusion away of stimulating substances produced by the marginal hyphae. Hawker (1954) reports that fruit-bodies of certain hypogeous fungi are frequently found in contact with tree roots, path edges or the hard pan of clay underlying the humus in woods. Here the contact stimulus, rather than the accumulation of any stimulatory substances, is probably the controlling factor.

LIGHT

The effect of light on sporulation of fungi is exceedingly complex. Both the nature of the response and the stage of development at which the fungus is light-sensitive vary with the quality and intensity of the light, with the duration of the exposure, with the species of fungus and with changes in other environmental factors. While there are numerous references to the effect of light on particular fungi, there are few intensive studies and still fewer attempts to interpret the chain of events leading from the light stimulus to the final manifestation of its effects. It is possible, however, to divide fungi roughly into four groups according to their response to light as follows: (i) those which are able to produce viable spores in complete darkness and which do not produce spores in greater numbers when illuminated; (ii) those which, while they are able to produce viable spores in complete darkness, do so more freely when illuminated at some

[1] See H. W. Buston and B. Rickard. *J. gen. Microbiol.* 15 (1956), 194–7.

time during development; (iii) those which are unable to produce viable spores in the complete absence of light; and (iv) those in which sporulation is actually checked or prevented by exposure to light at some stage in development. These groups are not always clearly separated and a fungus may come under one heading or another as a result of changes in environmental factors other than light, e.g. *Mucor flavidus* which, when growing on certain solid media, is able to form sporangia in the dark, but is unable to do so in a liquid medium unless illuminated (Lendner, 1897). Moreover, one type of spore of a particular fungus may be influenced by light, whereas another type is not, as with the formation of conidia and sporangia respectively by *Choanephora cucurbitarum* (Barnett and Lilly, 1950).

Fungi which sporulate equally freely in darkness or light are found among all the main taxonomic groups. The sporangia of many species of *Mucor* and related genera, the conidia of many Ascomycetes and Fungi Imperfecti, the perithecia of certain Pyrenomycetes and the fruit-bodies of some agarics show this independence of illumination, while representatives of the same families may be completely dependent on light for sporulation.

Many fungi, both in their natural habitat and in artificial culture, produce alternating zones of different growth type in response to diurnal changes in light intensity. Most of these examples of zonation are due to the effect of light in increasing the production of spores so that zones of intense sporulation alternate with those of sparse sporulation or even total absence of spores. Occasionally, however, zonation is due to the *inhibiting* effect of light on sporulation, e.g. *Sclerotinia fructicola*, Hall (1933), *Phyllosticta* sp., Stevens and Hall (1909), *Cephalothecium roseum*, Gallemaerts (1910) and Hedgecock (1906). Many species of *Fusarium* (Brown, 1925; Bisby, 1925; Snyder and Hansen, 1941) show zones of intensified production of macroconidia in response to light, while the brown-rot fungi, such as *Sclerotinia (Monilia) fructigena*, show conspicuous rings of dense masses of conidia, both on the rotting fruit and in culture.

The effect of light on *S. fructigena* has been investigated in detail by Hall (1933). Zonation of this fungus is shown clearly only in acid media, while *S. fructicola* produces rings over a wider range of pH. By analysis of the relation between periods of illumination and the position of zones of intensified sporulation, Hall shows that there is a lag period between the beginning of the light period and the visible increase in production of conidia. With long periods of illumination the formation of the zone of abundant conidia may begin before the end of the period, but with shorter exposures increased sporulation may take place entirely during the ensuing dark period. If a culture which has been exposed to regularly alternating light and darkness is transferred to continuous darkness, a zone of intense conidial production results from the last period of illumination, and one or more less clearly defined zones may be formed subsequently. The formation of such subsidiary ridges is difficult to interpret and one must conclude that some rhythm of branching is established by the regular alternation of light and darkness and that this rhythm is not immediately lost on returning to conditions of continuous darkness. Such a continuation of a rhythmic response after the cessation of a regular stimulus of alternating light and dark periods is also described by Ingold and Cox (1955) for the discharge of spores by fruit-bodies of *Daldinia concentrica*.

A similar stimulating effect of light on sporulation is probably the main factor determining the diurnal periodicity in spore development and discharge shown by many fungi in their natural habitats, e.g. *Pilobolus* spp., McVickar (1942), *Erysiphe polygoni*, certain downy mildews and *Taphrina deformans*, Yarwood (1936, 1937, 1941).

While zonation in artificial culture is usually the result of increased spore production as a result of periodic illumination, many fungi are unable to produce spores at all in complete darkness. Examples of such complete dependence on a period of illumination are to be found in most groups of fungi. As early as 1877, Brefeld showed that some species of *Coprinus* are unable to produce fruit-bodies in the dark. Exposure to light for 2 to 3 hours is sufficient to permit normal development to continue after the cultures are returned to conditions of complete darkness. Light is thus essential for the *initiation* of fruit-bodies of these species but, once formed, the initials continue to develop in the absence of light.

Results similar to these have been obtained by many workers for a wide range of fungi. Many Discomycetes fail to form apothecia in the absence of light. Among these *Pyronema confluens* was investigated by Robinson (1926). If this fungus is grown on a suitable medium in a petri dish it fills the dish within a few days and extension growth is then checked. This check is followed by the development of groups of short tufted branches. In complete darkness these eventually give rise to small black sclerotium-like bodies but no apothecia develop and the pink pigment associated with these is lacking. If at the time of the check to growth, or just before this, the cultures are exposed for 6 hours to a lamp of 40 candle-power at a distance of 50 cm. the tufts of short hyphae give rise to groups of antheridia and ascogonia, but no further development of these takes place in darkness. If, however, the culture is again illuminated or if the time of the original period of exposure to light is increased to 24 hours, normal apothecia, containing asci and viable ascospores, develop and the characteristic pink pigment is produced both in the fruit-bodies and in the surrounding hyphae. The apothecia mature first in the parts of the culture nearest to the light. Robinson concluded that some photochemical action of light is essential both for pigmentation and for the development of apothecia. He could find no difference in respiration rate between illuminated and non-illuminated colonies and the effect of light was not transmitted from an illuminated to a shaded part of the same culture.

While sporulation of many fungi is entirely dependent on light, with others the development of spores or fruit-body initials may be totally inhibited by light. Buller (1909–50) gives a detailed report of experiments with *Coprinus sterquilinus*, in which he shows that the action of light in preventing development of fruit-body rudiments formed in the dark is the chief factor in ensuring that only those fruit-bodies develop which are firmly 'rooted' at the base of the dung balls on which this fungus grows.

With many fungi, alternating periods of light and darkness are more favourable than continuous light, and in some examples it has been shown that *both* light *and* darkness are essential for normal development. Thus Barnett and Lilly (1950) found that a strain of *Choanephora cucurbitarum* was unable to form conidia in either continuous strong light or continuous darkness, while light had no effect on the formation of sporangia. Another strain described by Christenberry (1938) formed conidia in complete darkness but did so more freely with alternating periods of light and dark. Barnett and Lilly concluded that light influenced at least two metabolic reactions or groups of reactions, both of

41

which are essential steps leading to the formation of conidia. Light is essential to one of these (reaction A) but inhibits the second (reaction B) which is only able to occur in darkness or weak light. In continuous bright light only reaction A takes place, in continuous darkness only B continues, while in continuous weak light both reactions occur simultaneously. It is clear that the reactions leading to sporangial production must differ from those inducing conidial formation, at least in their sensitivity to light, and that in Christenberry's strain reaction B either differs from that in Barnett and Lilly's strain or can be replaced by some other mechanism which is lacking in the latter strain.

The value of alternating periods of light and darkness in the stimulation of fruiting is shown by Timnick *et al.* (1951) for ascospore production by *Diaporthe phaseolorum*. In continuous darkness a few perithecia containing numerous ascospores developed; in continuous light there were numerous perithecia but these contained few spores, while in alternating light and dark numerous perithecia containing numerous spores were formed.

Light not only plays a part in stimulating or inhibiting the initiation and development of spores and sporophores in many fungi, but it may influence their *rate* of growth or the relative rates of growth of different parts of the spore or sporophore, thus altering their morphology.

Even those fungi which do not show significant increases in sporulation with illumination may yet respond to light by an alteration in the growth rate of spore-bearing structures or in the rate of development of the spores. The effect of light on the growth of the sporangiophore and sporangium of *Phycomyces* spp. has been studied in detail by Blaauw (1914-19) and Castle (1928-9, 1929-30, 1930-1, 1931, 1931-2, 1932-3). Under constant conditions in darkness the sporangiophores of *Phycomyces*, which consist of a single cell, show a uniform increase in length over a period of 6 to 12 hours. When transferred to conditions of equal illumination on all sides these sporangiophores show a temporary increase in growth rate (Blaauw's 'light-growth response'). With high light intensity this initial increase in rate of growth is large, the reaction time (i.e. the time elapsing between the moment of exposure to light and the first measurable response) is short, and the increase in growth rate is followed by a decrease to a rate below that in constant darkness. The net result of long exposure to strong light is thus to decrease growth rate and final lengths of the sporangiophores. With weak light the reaction time is relatively long, the increase in growth rate is small and no subsequent depression of the rate below that in constant darkness is discernible.

Sensitivity to light decreases with prolonged exposure but is restored by a period in the dark. Castle (1931–2) concludes that in darkness a 'light-sensitive' substance accumulates in the sporangiophores and is activated by exposure to light, thus accounting for the phenomenon of 'dark adaptation'. After prolonged exposure to light this hypothetical substance would be exhausted and a further period of growth in the dark would be necessary before its accumulation would again make the sporangiophore light-sensitive.

Light is essential for the initiation of sporangia of *Pilobolus* (Page, 1956). After a subsequent period of darkness, elongation of the sporophore takes place on a second exposure to light in the presence of an adequate external supply of thiazole. Page concludes that the initial stimulus takes place through the absorption of light by a flavin and that during subsequent darkness a second (unknown) light-sensitive substance accumulates.

Light may also influence the final size and form of fruit-bodies or spores. Thus, whereas perithecia of *Melanospora destruens* (Asthana and Hawker, 1936) are equally numerous whether the cultures are illuminated or not, those formed by colonies exposed to diurnal illumination are about 50% larger than those in total darkness. Light also increases the relative length and degree of septation of macroconidia of some species of *Fusarium* (Harter, 1939; Snyder and Hansen, 1941).

While some Hymenomycetes, such as the commercial mushroom, are able to produce normal fruit-bodies in total darkness, others produce only abnormal ones. Pfeffer (1903) reports the abortion of rudimentary sporangiophores of *Pilobolus microsporus* and of fruit-body initials of species of *Coprinus* in the absence of light. Buller (1909) describes abnormal fruit-bodies lacking a pileus, or with only a rudimentary one, produced by *Polyporus squamosus*, *Lentinus lepideus*, etc., on naturally infected wood blocks in the dark. The development of such abnormal fruit-bodies by polypores in coal-mines is a frequent and well-known phenomenon. Plunkett (1956) shows that the pileus of

43

Polyporellus brumalis fails to form under conditions of high humidity and weak light, but either an increase in light intensity or a decrease in relative humidity allows the formation of normal caps. This work shows the importance of the interaction of different environmental factors.

It has been observed in this laboratory that, in the absence of light, a strain of *Thamnidium elegans* produces only sporangioles, but that in illuminated cultures both sporangioles and terminal multispored sporangia develop normally.

Some attention has been given to the effect of the wave-length of light on sporulation, but the results so far obtained are contradictory and difficult to interpret. Christenberry (1938) claimed that red-yellow light is the most effective in stimulating production of conidia by *Choanephora*, while blue light is said to be most effective in inducing the formation of apothecia by *Pyronema confluens* (Robinson, 1926), and in causing increased growth and consequent phototropic curvatures in the sporangiophores of *Phycomyces* spp. (Blaauw, *loc. cit.*; Castle, *loc. cit.*) and *Pilobolus* sp. (Parr, 1918). The last two fungi produce the orange pigment carotene and it has been suggested (Bünning, 1937) that carotenoids may be the 'light-sensitive' substance postulated by Castle (1931–2). Page (*loc. cit.*), however, has produced evidence that one of the light-sensitive substances in *Pilobolus* is a flavin (p. 43).

Ultra-violet light, in sub-lethal doses, has been shown to stimulate sporulation of some fungi, but is highly lethal in anything but very small amounts (Stevens, 1928). The specific effects on sporulation of other forms of radiation, outside the visible spectrum, have not been investigated.

The duration of the light stimulus required is often very short. Bisby (1925) obtained the formation of conidia of *Fusarium discolor sulphureum* by an exposure of $\frac{1}{4}$ second to bright daylight, while none formed in total darkness.

It seems most likely from a consideration of the available data that light must act by supplying energy leading to a photochemical reaction which is an essential step in the complicated chain of metabolic processes involved in spore formation. It is clear that this reaction can be performed by some fungi in the absence of light and that with many the effect of light is replaceable wholly or in part by some other factor, e.g. *Polyporellus brumalis* (see p. 46), where humidity and light are interacting

44

factors; *Helminthosporium gramineum* (Houston and Oswald, 1946), where conidia are formed in darkness on infected barley leaves but not on potato-dextrose agar; *Sphaerographium fraxini* and some other members of the Sphaeropsidales (Leonian, 1924), which produce more pycnidia in the dark at 30° C. than at the temperature at which they are produced freely in the light, namely room temperature.

Coons (1916) considered that the effect of light might be replaceable by certain oxidizing agents. He was able to induce the formation of a few pycnidia of *Plenodomus fuscomaculans* in the dark in cultures treated with hydrogen peroxide and some other oxidizing substances, but it was clear from his experiments that the physiological condition of the cultures at the time of the addition of these substances was critical. The amount of light which would induce the formation of pycnidia varied with age of the cultures. The maximum effect of hydrogen peroxide occurred when the cultures were in the condition when 1 hour's exposure to light would have been effective. Further critical studies of particular fungi in relation to light are highly desirable. Attention to the nature and function of the pigments which are produced only in illuminated mycelia and which are then often associated with intensified sporulation would also be of value.

GRAVITY

While vegetative hyphae are usually ageotropic, reproductive structures are often strongly geotropic. Gravity plays a part as important as that of light in determining the final orientation of the spore-bearing structures of many fungi. Nothing is known of the mechanism involved. The effects of gravity are of particular importance in spore-dispersal mechanisms.

INTERACTION OF ENVIRONMENTAL FACTORS

It is clear from this account of the effects of environmental factors that the action of any one of them may be profoundly modified by changes in the others. Too few of the investigations of the effect of a particular factor take this into consideration. Recent studies on *Choanephora cucurbitarum* and on certain Basidiomycetes are notable exceptions and have consequently achieved a more complete picture of the effect of the environment on sporulation. From such studies in turn we may hope to

45

gain an insight into the fundamental metabolic changes within the mycelium, which lead to the change from the vegetative to the reproductive phase.

It must be supposed that reproduction, particularly where the formation of large and complex fruit-bodies is concerned, requires the synthesis of a wider diversity of enzymes and other complex substances than those sufficient for vegetative growth, and that this in turn demands the release of relatively more energy. It is therefore not surprising to find that the environmental factors permitting sporulation are often more restricted in range than for mycelial growth. The fact that, in general, the minimum temperature necessary for sporulation is higher than that for growth, may be explained by the well-known effect of temperature on the rate of chemical reactions. At very low temperatures, essential reactions may either be entirely inhibited, or may fail to produce some necessary product or intermediate substance in sufficient quantity for spore production, before other factors, such as staling of the medium, prevent further development. The effect of humidity is complex and may act in more than one way. Spore-bearing structures are often aerial, and are thus more exposed to the desiccating effect of the atmosphere than are the vegetative hyphae on the surface of or within the substratum. On the other hand the translocation of large quantities of food materials into such a structure as a developing sporangium of *Mucor* or the conidial head of an *Aspergillus* almost certainly depends, at least in part, on the transpiration pull and hence many such structures fail to develop with too high a relative humidity.

As will be seen in the next chapter, sporulation is often preceded by a temporary intensification of respiration with consequent release of energy. The effect of poor aeration and of certain poisons may well act through inhibition of respiration. The importance of pH may also be largely due to its influence on respiration. It has already been pointed out that light most probably acts by supplying energy for a photochemical reaction or group of reactions. With some fungi this energy source may be replaced to some extent by others, for example certain nutrients or increased temperature. The replacement of bright light by low humidity as a factor in normal development of the pileus of *Polyporellus brumalis* is more difficult to interpret, although one might suggest that lower humidity stimulates the

46

translocation of larger quantities of sap to the stipe apex, and may thus provide sufficient of some necessary substance which is synthesized more slowly in weak than in strong light. Thus, with strong light and high humidity a relatively small quantity of sap, rich in the particular substance, would reach the stipe apex, while with lower humidity and weak light the sap would be poorer in that substance but, by arriving in greater quantity, would provide the necessary threshold amount. The view that spore production depends on energy relations, and the consequent synthesis of diverse complex substances, is supported by the study of the effects of nutrition, which will be considered in the next chapter.

CHAPTER 5

THE EFFECT OF NUTRITION ON SPORULATION

NUTRITION is perhaps the most important single factor in the control of growth and reproduction. The influence of type and quantity of food supply on sporulation has been more intensively studied than that of any other factor. The general food requirements of fungi are now well known. This information has been summarized by Hawker (1950) and by Lilly and Barnett (1951). Owing to their lack of chlorophyll, fungi are incapable of photosynthesis of carbohydrates from carbon dioxide and water, nor do they include any forms comparable with the autotrophic bacteria, which are able to carry out this synthesis with the aid of chemical energy derived from the oxidation of inorganic compounds. Thus they must all be supplied with suitable organic material in the substrate and the most generally suitable substance is glucose. In common with other organisms they must also be supplied with nitrogen. No species of fungus has been proved to fix nitrogen, although it has been suggested that certain species, notably mycorrhizal forms, may be able to do so under specific conditions. Their requirements, however, vary from simple nitrates or ammonium salts to complex organic nitrogenous compounds. In addition to carbohydrates (or a limited range of other carbon compounds) and nitrogen, fungi must be supplied with phosphorus, potassium, magnesium and sulphur in measurable quantities. It has also been demonstrated that a number of metals, such as iron, zinc, copper, manganese, etc., are required in such minute quantities that they are termed 'trace elements'. In addition to these essential nutrients, which are required by all fungi, traces of complex organic susbstances are necessary for growth of certain species, or strains, which are unable to synthesize them from simple carbohydrates and salts. Both trace elements and these organic 'growth-substances' have been shown to be, or are suspected to be, essential components of various enzyme systems.

The nature, the concentration and the total quantity of the

48

food supply, all play a part in determining the time of initiation of spores or sporophores, the quantity of these produced and, to some extent, their form and size.

The mycelium must obviously accumulate food material and synthesize from it sufficient complex substances (polysaccharides, proteins, fats, enzymes, etc.) before spores can be formed or fruit-bodies initiated. With some types of spore, such as many asexual conidia, the food requirements are not much greater than those for mycelial growth, so that these spores may develop early in the life of a particular colony and on a mycelium which has not been richly fed. With other spores, produced on or in complex fruit-bodies, a relatively enormous reserve of food must be amassed and much energy expended in transforming this into more complex substances before such a fruit-body can develop, and this activity must be maintained until the spores are mature. The distribution of mature fruit-bodies in cultures of *Coprinus lagopus* is determined by the flow of food into those developing first, with a resulting loss of dry weight by the mycelium (Madelin, 1956).

In his classic studies on reproduction in fungi, Klebs (1898, 1899, 1900) showed that strains of the water mould, *Saprolegnia*, remained in the vegetative condition in crude culture for long periods if kept supplied with fresh nutrient, but produced sporangia within a few days of transfer to water. His results were confirmed by later workers (Kauffman, 1908; Kanouse, 1932) and similar results have been obtained with some other fungi. More commonly, however, such a drastic reduction in food supply as takes place upon transfer to water, prevents or reduces sporulation of even a well-nourished mycelium. Transfer from a rich to a more dilute medium is less drastic and is often effective in stimulating reproduction, as with some members of the Sphaeropsidales (Leonian, 1924). Sporulation may also take place freely on young hyphae which are allowed to grow on to a dilute medium while maintaining organic connexion with the parent mycelium growing on a rich one. Thus Claussen (1912) showed that apothecia of *Pyronema confluens* are produced in a ring where hyphae grow over the edge of a small petri dish containing an inulin medium on to one lacking

49

inulin. Similarly, Asthana and Hawker (1936) showed that, if a segment of a glucose-salts agar plate, on which *Melanospora destruens* is growing, is removed and replaced by one of plain agar or dilute glucose-salts medium, perithecia form earlier and in greater number in this replacement segment.

There is little doubt that sporulation under natural conditions is often induced by a fall in the level of food materials. Many leaf-inhabiting fungi, such as certain downy mildews, *Venturia inaequalis*, *V. pyrina* (apple and pear scab), *Gnomonia erythrostroma* (cherry leaf scorch), etc., produce conidia on the living leaves, but do not form their perfect stages until the leaves are dead or dying. Many wood-destroying polypores remain in the vegetative state in the living tree for long periods and produce their fruit-bodies only when the tree is in a semi-moribund condition. In all these examples the reduction in food supply which presumably leads to sporulation is preceded by a period of intensive feeding under conditions of high availability of nutrients. Similarly, the production of conidia by the predacious fungus, *Arthrobotrys oligospora*, is 'usually preceded by an initial feeding period during which the young mycelium gorges itself on eelworms' (Duddington, 1955). Such a preliminary feeding period is to a greater or lesser extent essential before sporulation is possible.

In artificial culture a reduction in food supply must obviously occur as a result of growth on a finite quantity of medium. This in itself may often be sufficient to induce sporulation, but on an initially rich medium the products of metabolism may accumulate to an extent which inhibits reproduction and eventually growth itself. If this happens before the food supply has dropped to a level favourable for sporulation, no spores can develop and the culture remains in a vegetative condition. Such an interplay of factors explains the somewhat conflicting results obtained by different workers who have studied the effect of initial concentration of food on fungal growth and reproduction. In general the range of initial concentration of nutrients which permits asexual sporulation is narrower than that allowing mycelial growth, and that for sexual reproduction is often still more limited.

The requirements of the ascomycete, *Melanospora destruens*, have been worked out in detail (Hawker, 1936, 1939; Hawker and Chaudhuri, 1946). When grown on a glucose-salts medium this fungus produces perithecia fairly freely when the initial concentration of glucose does not exceed 0·5%, but with increasing concentration the number formed falls rapidly to zero. If, however, the level of glucose is maintained at or near 0·5% by the addition of small amounts to the cultures daily or by repeated transfer of the mycelium to fresh medium at that concentration, the number of perithecia formed is greatly increased. This indicates that the *amount* of glucose supplied in a medium with an initial concentration of 0·5% is below the optimal amount for the production of perithecia but that, with higher initial concentrations, other factors, such as the accumulation of toxic metabolic products (staling factors), are limiting.

Plunkett (1953), working with *Collybia velutipes*, showed that in a liquid medium containing sucrose, asparagine and salts, increase in asparagine concentration soon limits fruiting, but that the total weight of sporophores increases with an increase in sucrose concentration. Mycelial growth continues at concentrations of asparagine above that which inhibited fruiting. Some of these effects were attributed to unfavourable changes in pH. Low initial concentration of sugar causes an earlier formation of fruit-body initials which do not always mature. He supplied large quantities of relatively weak medium to mycelia of *Collybia velutipes* and showed that while a high initial concentration is unfavourable, a high total quantity of food material is essential for the formation and maturation of the maximum number of fruit-bodies.

Thus high initial concentration of food material is probably not in itself harmful, and may even be beneficial for a short period, but leads to vigorous vegetative growth which in turn tends to modify the medium, by changing its pH value or by the production of actual toxic staling products, in such a way that the delicate metabolic balance leading to reproduction is upset. Much of the difficulty experienced in inducing the production of the complex fruit-bodies of the Higher Fungi in artificial cultures may be due to failure to provide sufficient food materials without the deleterious effects which follow rapid growth in a small quantity of concentrated medium. A continuous supply, or an initially large quantity of a relatively dilute medium, is much more likely to produce conditions suitable for the growth and maturation of fruit-bodies than is the provision of a small quantity of a rich medium.

Concentration of the medium may control the type of spore produced or the morphology of spores or spore-bearing structures. Thus both *Phycomyces blakesleeanus* (Leonian and Lilly, 1940) and *Sporodinia grandis* (Baker, 1931) require more carbohydrate for zygospore formation and maturation than for the production of sporangia. Goldring (1936) has shown that with low food supply *Blakeslea trispora* produces only sporangia, but that with increased concentration of nutrients both sporangia and sporangioles are produced. When *Thamnidium elegans* is grown for several generations under conditions of good nutrition, the sporangioles, which are usually small and contain few spores, become as large as, and contain as many spores as, the terminal multispored sporangium (Bachmann, 1895). Conversely, under conditions of poor nutrition, the number of spores in the terminal sporangium may be finally reduced to one. The conidia of *Cephalothecium gregalium* are normally uniseptate, but become abnormally elongated and form additional septa when grown on water agar (Hamilton and Boosalis, 1955). The terminal cell may even become abstricted.

THE NATURE OF THE FOOD SUPPLY

(a) *Source of carbon.* The nature of the food supply is also important in influencing reproduction. Since organic compounds are the sole sources of energy utilized by fungi, it is not surprising that carbon compounds play a large part in determining morphogenesis. Just as the initial concentration of food material favouring sporulation seldom corresponds to that optimal for vegetative growth, so the type of carbon source giving maximum growth is seldom the most suitable for spore production.

Melanospora destruens (Hawker, 1939) provides an example of the effect of carbon source on the type of growth. The number of perithecia produced by this fungus is greatly influenced by the nature and concentration of carbohydrate or other carbon compound present. As already described (p. 51), when glucose is the sole source of carbon, fruiting falls off rapidly as the concentration increases above 0·5% and is entirely inhibited at 2%. Vegetative growth, however, continues to increase until a concentration of about 10% is reached, when osmotic factors become limiting. Fructose or a mixture of glucose and fructose gives similar results. The effect of sucrose as sole source of carbon is, however, entirely different. Both growth and fruiting increase with increasing concentration of this sugar up to about 10% and at this value the number of perithecia produced is greater than that produced at the optimum concentration of glucose. Other carbohydrates, such as starch, show an effect intermediate between those of glucose and sucrose (Table 3). Some other Ascomycetes were investigated by Hawker and Chaudhuri (1946). These show the same kind of response to hexose sugars but differ in their response to various di- and poly-saccharides. It was shown that the type of response depends upon the rate at which a particular carbohydrate is hydrolysed by a particular fungus. The concentration of hexose sugars in the medium is thus an important factor. In extreme cases hydrolysis is so slow that a starvation type of growth with poor fruiting results at all concentrations (see fig. 3). It has already been pointed out that hexose sugars favour fruiting of *M. destruens* if they are given in small increments. In one experiment these were so adjusted as to reproduce as closely as possible the conditions in a medium originally containing 5% sucrose, but perithecial production was still inferior to that on the sucrose medium.

TABLE 3. *Effects of various sources of carbon and their concentrations on the formation of perithecia by* Melanospora destruens. (Data from Hawker, 1939.)

Source of carbon	Amount of carbon compound g./100 ml.						
	0·0	0·5	1·0	2·0	5·0	10·0	20·0
Glucose	3·0	7·2	3·2	Few	0	0	0
Fructose	3·0	8·3	3·0	Few	0	—	—
Maltose	3·0	8·2	5·4	4·7	—	—	—
Arabinose	3·0	10·0	11·8	7·2	—	—	—
Lactose	3·0	8·8	11·3	10·2	9·8	7·0	—
Starch	3·0	7·7	11·1	13·1	10·2	7·0	—
Sucrose	3·0	2·9	3·2	5·2	9·4	9·9	Few
Galactose	3·0	3·5	1·6	1·0	—	—	—
Inulin	3·0	3·5	3·7	2·0	—	—	—
Raffinose	—	—	2·3	4·6	—	—	—
Mannitol	3·0	3·4	3·2	3·4	3·6	0	—

Basal medium contained: KNO_3, 3·5 g.; KH_2PO_4, 1·75 g.; $MgSO_4$, 0·75 g.; lentil concentrate as source of growth-substances, 0·5 cc.; distilled water, 1 litre; agar, 15 g. Small-scale experiments in which pure biotin and thiamin were used instead of lentil concentrate gave comparable results. Figures are mean numbers of perithecia per microscopic field in ten counts on each of three to ten plates. A dash (—) means no data.

53

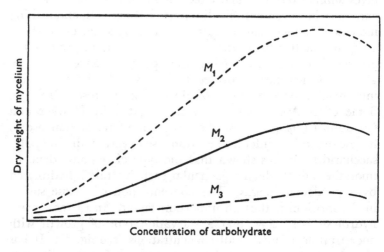

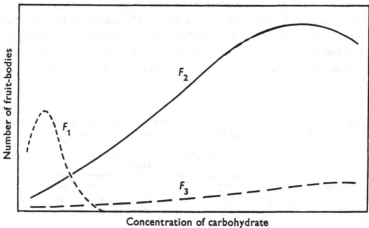

Fig. 3. Effect of concentration of carbohydrate on growth and fruiting of certain ascomycetous fungi. The curves show three types of response to concentration of carbohydrate. *M* curves show the effect on mycelial growth, *F* curves show effect on fruiting. M_1 and F_1 illustrate the typical response to hexose sugars (as with *Melanospora destruens* on glucose). M_2 and F_2 represent the effect of a more complex carbohydrate on a fungus able to break it down at a moderate rate (as with *M. destruens* on sucrose). M_3 and F_3 show the extreme type of response where the ability to break down a particular carbohydrate is so poor that a starvation type of growth results (as with *Pyronema confluens* on lactose). In contrast, when a fungus is able to break down such a complex carbohydrate rapidly the result is that shown by curves M_1 and F_1, i.e. equivalent to growth on a hexose sugar (as with *P. confluens* on sucrose) (after Hawker and Chaudhuri, 1946).

It was concluded that some intermediate substance, produced during inversion of sucrose, might favour fruiting. This view was supported by the observation that factors tending to a slight increase in rate of inversion of sucrose, such as increase in temperature, or the presence of a weak solution of invertase, or of a colony of a fungus able to invert sucrose more rapidly than *Melanospora destruens* itself, all tended to an increase in perithecial production on a sucrose medium (Hawker, 1947a). The similarity in effect of raffinose and sucrose suggested that furanose fructose might be of significance. Inulin, which gives fructose on hydrolysis, is not used by *M. destruens*, but if this fungus is grown on an inulin medium in mixed cultures with another organism capable of hydrolysing this carbohydrate the effect on fruiting is favourable. The good effect of arabinose (see Table 3) might be attributed to the structural similarity of arabo-furanose and fructo-furanose. The higher alcohol, mannitol, has some structural resemblance to fructose but lacks the ring structure. *M. destruens* cannot use mannitol but, in the presence of other organisms which can use this substance, vegetative growth takes place, although no perithecia are formed, a circumstance suggesting that mannitol is not normally used. It is clear that the ring structure of fructo-furanose is the important factor.

In the utilization of sucrose by sugar-cane leaves and yeast, fructose diphosphate is formed, and it was thought probable that this ester might be favourable to perithecial formation. This, however, would not explain the superiority of starch, maltose, and lactose over hexose sugars. Doudoroff *et al.* (1943) showed that the bacterium *Pseudomonas saccharophila* is able to synthesize sucrose from a mixture of fructose and glucose-1-phosphate. This ester is also concerned in the synthesis of starch and glycogen in higher plants and animals. Glycogen itself is a favourable source of carbon for fruiting of *Melanospora destruens*. It was therefore concluded that the ease of phosphorylation of various carbohydrates is one of the critical factors in determining their effect on fruiting. Glucose-1-phosphate and fructose diphosphate may perhaps be readily interconverted and the ease with which either of them is produced may govern the metabolic chain leading to the production of perithecia. Small quantities of fructose diphosphate, glucose-1-phosphate or of a mixture of hexose monophosphates were added to both

55

glucose and sucrose media and gave increased fruiting (Hawker, 1947 a, 1948). Larger quantities depressed sporulation.

Buston *et al.* (1953), in an investigation of the favourable effect of jute extract on the production of perithecia by *Chaetomium globosum*, demonstrated by paper chromatography that the extract contained traces of glucose-1-phosphate and fructose-1 : 6-diphosphate. They further showed that these substances are partly responsible for the favourable effect of the extract on fruiting. In later work (Buston and Khan, 1956) a number of strains of *Aspergillus fumigatus* were examined for their effect on *Chaetomium*. Some of these stimulate fruiting and others have little or no effect. Mycelia of all strains contain organic phosphate, some of which has been identified as phosphoglyceric acid, but only those which excrete this substance into the medium have a stimulating effect on *Chaetomium*. This is of particular interest in view of Bretzloff's (1954) inference that since *Neurospora* cell walls are impermeable to phosphate esters all fungi must have similarly impermeable walls.

Buston's work demonstrates that the permeability to organic phosphate substances may vary even in strains of the same species. It shows, moreover, that some fungi are able to produce them within the mycelium. It may well prove to be the inability to do this that makes *Melanospora* and *Chaetomium* so responsive to an external supply of hexose phosphates and phosphoglyceric acid respectively. Small quantities of glucose-1-phosphate or fructose-1 : 6-diphosphate, when added to a suitable medium, stimulate the production of ascospores by the yeast *Saccharomyces cerevisiae* also (Miller *et al.* 1955). There seems to be little doubt that these substances do play a part in sporulation but the nature of their influence is at present obscure. It is significant, however, that the rate of respiration of *Melanospora destruens* on sucrose, raffinose and other favourable carbohydrates, or in the presence of traces of glucose-1-phosphate is higher than on a glucose medium (Hawker, unpublished data). The effect of various carbohydrates on production of conidia by *Fusarium fructigenum* is also correlated with respiration rate (Brown, 1925). It is likely that the effect on respiration rate, and hence on the available energy for chemical synthesis, is the important factor.

Recent studies on the Basidiomycetes give somewhat contradictory results. Thus Plunkett (1953) states that sucrose and glucose are indistinguishable in their effects on growth and fruiting of *Collybia velutipes*, while Aschan (1954) claims that, for her strain of this fungus, glucose is slightly superior to sucrose, which in turn is superior to maltose. This apparent slight discrepancy may be due to the different rate of inversion of sucrose by different strains of the fungus or to the different sources of nitrogen used, i.e. asparagine and ammonium tartrate respectively. Bille-Hansen (1953), in a preliminary investigation of three species of *Coprinus*, found that glucose caused good growth in all these, but that only *C. congregatus* produced fruit-bodies. Sucrose produced poor growth in all species and only dwarf fruit-bodies developed, but with maltose all grew well and produced normal fruit-bodies. No attempt was made to estimate the ability of any of these fungi to invert sucrose or hydrolyse maltose, but the results suggest that the rate of breakdown of these sugars may determine their effects, as with the Ascomycetes studied by Hawker and Chaudhuri (*loc. cit.*). It is clear that, as would be expected, larger total amounts of sugar are necessary for the maturation of the large fruit-bodies of the agarics than for the small perithecia of the Pyrenomycetes.

Examples of the differing responses of various fungi to different carbohydrates are numerous, but few attempts have been made to explain these differences. It is probable that, with some complex carbohydrates, purification is insufficient to remove all traces of growth-substances (see p. 64), but it is likely that rate of hydrolysis of complex carbohydrates and the nature of the intermediates produced are of importance, as with *Melanospora destruens*. The frequent superiority of 'natural' media over synthetic ones may be in part owing to the presence of suitable poly- or disaccharides in the former.

Carbon compounds, other than carbohydrates, such as the higher alcohols, organic acids and organic nitrogen substances are not usually particularly suitable for spore production. This is almost certainly owing either to the inability of the fungus to use them at all or, when they are used, to the rapid development of unfavourable pH values in the medium.

(*b*) *Nitrogen source and carbon-nitrogen ratio.* Detailed studies of the effect of the nature and concentration of nitrogen compounds on sporulation in fungi are surprisingly few. In view of the great interest of nitrogen metabolism and the well-known effect of carbon-nitrogen ratio on reproduction in other organisms, notably the higher plants, one would expect investigations of the effect of nitrogen on fungal reproduction to yield results of comparable interest.

The minimum amount of nitrogen permitting sporulation is usually above that at which sparse vegetative growth can take place. If the amount is increased much above this minimum value, however, vegetative growth becomes very vigorous and sporulation is often inhibited. This may be due to exhaustion of other essential nutrients by the growing mycelium, or to the rapid accumulation of staling substances. Nitrogen compounds may, however, be themselves inhibitory, and it has been shown in several examples that nitrogen or some particular nitrogenous compound must be nearly or entirely used up before the initiation, or sometimes the maturation of fruit-bodies, can take place. Robinson (1926) showed that, while a minimum initial concentration of nitrogen in the medium is essential for growth and apothecial production by *Pyronema confluens*, the reproductive process is initiated only after almost complete exhaustion of the nitrate ion. Westergaard and Mitchell (1947) noted that the carbon-nitrogen ratio is of importance in the fertilization process in *Neurospora crassa* and that a high concentration of casein hydrolysate inhibits fruiting.

Hirsch (1954) investigated the effects of inorganic and organic nitrogen on this fungus in more detail. His results are summarized in Table 4. Formation of protoperithecia reaches a maximum at a concentration of $0 \cdot 1 \%$ potassium nitrate. Growth continues up to the highest concentration used (8%) but initiation of protoperithecia is inhibited. Complete exhaustion of nitrate is not a prerequisite for formation of protoperithecia but must precede their development into mature perithecia. An occasional perithecium may develop in the presence of nitrate but never produces viable ascospores. Hirsch concludes that the nitrate ion is a specific inhibitor of some phase in the sexual cycle. This agrees with Robinson's results cited above, and with Leonian and Lilly's (1940) finding, that the addition of ammonium nitrate to a medium containing amino acids prevents the formation of zygospores by *Phycomyces blakesleeanus*. Hirsch further shows that formation of protoperithecia is inhibited by high concentration of organic nitrogen in the form of hydrolysed casein. This is not due to increased buffering capacity nor to an accumulation of nitrate. When a mycelium is grown on a medium with a high nitrogen content and is

58

TABLE 4. *The effect of nitrate and organic nitrogen on the production of perithecia by Neurospora crassa.* (Data from Hirsch, 1954.)

Time (days approx.)	Nitrogen source KNO_3 (g./l.), casein hydrolysate (ml./l.)							
		0·1 KNO_3	1·0 KNO_3	2·0 KNO_3	8·0 KNO_3	1·0 KNO_3 5 ml. C.H.	1·0 KNO_3 50 ml. C.H.	1·0 KNO_3 150 ml. C.H.
3	0	0·4	0–2·5	0	0	0	0	0
5	0	0·9	21–23·0	0·4	1·6	15·5	0	0
13–14	0·3	10·0	48–43·0	11·6	1·6	46·3	27·7	0
Test for nitrate in medium	—	—	—	—	+	—	+	+
Perithecia	+	+	+	+	—	Pseudo-perithecia	—	—

c.h. = casein hydrolysate. Basal medium contained: KH_2PO_4, 1·0 g.; $MgSO_4$, 0·5 g.; NaCl, 0·1 g.; $CaCl_2$, 0·1 g.; biotin, 10 μg.; trace element solution, 1 ml.; glycerol, 20 ml.; distilled water, 1000 ml.; agar, 15 g. The second figures in the fourth column are controls to those in columns 7, 8, and 9. The figures represent mean numbers of protoperithecia from counts on five fields on each of two plates. Plates were tested for nitrate after 13–21 days, conidiated with a suitable strain, incubated at 25° C. for a further 14 days and finally scored for presence of mature perithecia and ascospores.

59

transferred to one with a lower nitrogen content, protoperithecia are formed, showing that the high nitrogen medium did not destroy the capacity of the fungus to reproduce.

Ronsdorf (1931), Schopfer (1931, 1931a) and Robbins (1939) showed that high concentration of asparagine inhibits zygo-spore formation of *Phycomyces blakesleeanus*, while at the same time increasing vegetative growth. Robbins showed that the amount of asparagine necessary to inhibit the formation of progame-tangia is greater than that sufficient to prevent the formation of mature zygospores from these. He suggested that the vigorous vegetative growth with high concentrations of asparagine leads to the exhaustion of the available growth-substances in the medium, thus preventing reproduction. The addition of certain plant extracts, which presumably contain growth-substances, reduces the inhibiting effect of asparagine. Schopfer demon-strated that the addition of gelatine to an otherwise suitable medium, such as malt extract, also prevents zygospore produc-tion. Plunkett (1953) also records an inhibitory effect of high concentration of asparagine on the production of sporophores by *Collybia velutipes*, while Bille-Hansen (1953, 1953a) shows that *Coprinus sassii* fails to fruit on an asparagine medium. The fact that the utilization of asparagine by many fungi in the presence of a low concentration of carbohydrate tends to the release of free ammonia, and the consequent development of a high alkalinity, is probably one reason for the harmful effects of asparagine on fruiting. The general tendency for such harm-ful changes in pH to develop in media of a low carbon-nitrogen ratio (Brown, 1923) may explain claims made that in some instances a high ratio is beneficial.

The carbon-nitrogen ratio may also influence the type and the form of spores produced. Peiris (1947) showed that the production of conidia by *Botrytis cinerea* is favoured by a rela-tively high ratio of peptone to glucose, and inhibited by high concentrations of glucose which favour sclerotial production. Townsend (1952 and in the press) confirmed this result and obtained similar ones when the source of nitrogen was aspara-gine, nitrate or ammonium salts.

High concentration of nitrogen reduces the number of septa in macroconidia of species of *Fusarium* (Brown and Horne, 1926; Horne and Mitter, 1927). Similarly the shape and size of conidia of species of *Cytosporina*, *Phomopsis* and *Diaporthe* vary from small and oval to large and curved or even elongated and sickle-shaped, as the proportion of glucose to asparagine is increased (Nitimargi, 1937).

Occasional claims that organic forms of nitrogen, such as peptone, are superior to inorganic compounds in inducing fruiting have usually been found to be due to the presence of traces of vitamins in the organic compounds. There is no foundation for the belief that such organic compounds are uniformly superior to inorganic ones as sources of nitrogen. When a strain has lost the ability to synthesize one or more essential amino acids (as with certain artificially induced mutants of *Neurospora* spp.), these must be supplied to permit growth and are obviously also essential for reproduction. While the form of nitrogen favouring growth is usually also good for sporulation, Subramanian and Pai (1953) showed that growth of *Fusarium vasinfectum* is best with potassium nitrate as the source of nitrogen, but that sporulation is better when urea is substituted for the nitrate. A urea medium has less residual nitrogen left after growth of this fungus than have nitrate or ammonium salt media and this may be the important factor.

It is clear that the effect of nitrogen in various forms on reproduction requires much more critical study than it has yet received.

(*c*) *Phosphorus, potassium, magnesium and sulphur.* These are all essential elements for both growth and sporulation and are required in the same form for both. Thus, in artificial media, phosphorus is usually supplied as a phosphate, but organic phosphorus in some forms can also be used. Potassium is usually supplied as potassium phosphate, potassium nitrate or both these. Magnesium and sulphur are usually supplied as magnesium sulphate, of which only a small amount is required. Fungi can use a variety of organic sulphur compounds but sulphites are toxic to most species.

As with carbon and nitrogen, these elements are required at a slightly higher level for spore production than the minimum permitting vegetative growth. Thus sparse mycelial growth takes place when *Fusarium fructigenum*, or certain other species of *Fusarium*, are grown in Richard's solution lacking added magnesium sulphate, but no conidia are formed. The traces of magnesium present as impurities in the other chemicals used and in the inoculum thus suffice for a minimum amount of

61

mycelium to develop but are insufficient for conidial production. Decrease in potassium similarly has a greater influence on sporulation than on vegetative growth.

Phosphorus plays a part in the initial breakdown of sugars, through the process of phosphorylation. It has already been shown (pp. 55, 56) that certain hexose phosphates favour spore production by certain fungi. The amount of phosphorus required is small, as it is not used up in the process but repeatedly becomes available for further phosphorylation of sugars. Nevertheless in active sugar breakdown, such as has been shown to precede sporulation, the amount required is by no means negligible. Since phosphorus enters into the composition of phospholipids and of nucleic acids, it will obviously be required in greater amounts during the active reproductive phase than for the maintenance of hyphal growth.

Since sulphur is present in all proteins, it is obviously essential for all growth, and the concentration could become limiting in the production of complex sporophores, or of spores with particularly dense cytoplasmic contents.

(d) *Calcium.* Calcium has usually been considered a non-essential element for fungal growth. Calcium carbonate is often used in artificial media to prevent the accumulation of excess acid and thus may be favourable to spore production. Recently, however, Basu (1951) showed that calcium is essential for the optimal formation of perithecia by *Chaetomium globosum* and increases fruiting, vegetative growth, or both, of several other species of *Chaetomium*. It can be replaced, in part at least, by strontium or barium. He claimed that the presence of calcium in jute extract largely accounts for the beneficial effect of the latter. In his experiments calcium was supplied in the form of calcium chloride. He suggested that more stringent measures to free glassware, chemicals and inoculum from calcium might also demonstrate the need of traces of this element for vegetative growth. Buston *et al.* (1953) confirmed that the calcium ion is required for perithecial formation in *C. globosum* but showed that the favourable effect of jute extract is due to a combination of the effects of calcium, sugar phosphates (see p. 56) and some other unidentified factor or factors.

Brian and Hemming (1950) demonstrated an increase in the number of conidia produced by certain strains of *Trichoderma viride* on the addition to the medium of calcium chloride

(0·01–0·1%). Only one strain responded similarly to lithium chloride.

(*e*) *Trace elements.* Minute quantities of certain heavy metals, such as iron, zinc, manganese, copper and molybdenum, and probably gallium and some others, are known to be essential for growth of fungi. *Aspergillus niger* has been most studied in this connexion. The need for iron and zinc was first recognized by Raulin (1869), but it was not until Steinberg (1919) developed better methods of freeing glassware, chemicals and inoculum from traces of metals, that the need for the majority of these was demonstrated. Many of them, such as copper, are required in such minute quantities that it is difficult to demonstrate that they are actually essential. Since sporulation usually requires more of these trace elements than does vegetative growth, it is easier to demonstrate this need. It is known that these metals enter into the composition of certain spore pigments and form essential parts of various important enzyme complexes, so that their essential role in growth and sporulation is obvious.

Thus it is usual for growth to be prevented in the absence of, or at very low concentrations of, a particular trace element. At a level which just permits mycelial growth, spore production may be inhibited. With slightly higher concentrations spores will develop, but the range of concentration allowing sporulation is usually rather narrow.

Concentrations of copper just sufficient for the production of a few conidia of *Aspergillus niger* prevent pigmentation of the spores. With increase in concentration from 0·05 to 1·0 µg. per 50 ml. medium the colour changes from yellow through brown to the normal black. This relation between copper concentration and spore colour provides a sensitive assay method for this metal (Mulder, 1938). Polyphenol oxidase activity of the spores is proportional to the amount of available copper in the culture medium and is at a maximum at 1·0 µg. Cu per 50 ml.

The formation of carotenoid pigments in the conidia of certain strains of *Neurospora* is prevented by deficiencies in any of a number of trace elements, including iron, copper, zinc, manganese or molybdenum. *Excess* zinc also inhibits pigmentation and increased zinc in the medium is used to suppress pigment formation when the felts are grown for use in enzyme assays (personal communication from Dr D. J. D. Nicholas, to

63

whom the author is indebted for much of the information given in this section).

Excess zinc readily suppresses spore formation, e.g. conidia of *Aspergillus niger* (Porges, 1932; Roberg, 1928). Absence of zinc inhibits the production of spores, as with the conidia of *Fusarium oxysporum* and *Acaulium nigrum* and of zygospores of *Rhizopus nigricans* (Niethammer, 1938).

There is little carry-over of trace metals in spores. A single spore of *Aspergillus niger* is said to contain only $1 \times 10^{-50} \mu$g. of molybdenum. Spores from felts grown on media deficient in any of the effective trace metals are viable and germinate well when transferred to a complete culture medium.

The requirements of different strains of the same species may differ widely. The Steinberg strain of *A. niger* is able to form a few conidia at levels of various trace metals below those which entirely inhibit conidium production by the Mulder strain.

(*f*) *Growth-substances.* Growth-substances include a number of complex organic compounds which are essential for the growth of living organisms but which are required in very small quantities. Many fungi are unable to grow without an external supply of one or more of such substances and are said to be 'deficient for' these. Others, which are independent of an external supply, have been shown to be able to synthesize the necessary growth-substances within the hyphae.

Growth-substances needed by various fungi include thiamin (vitamin B_1), biotin (vitamin H), pyridoxine (vitamin B_6), certain purines (including adenine, guanine, hypoxanthine), inositol, which, however, is required in larger quantity than a typical growth-substance, and a number of unidentified factors. Most of these are known to play a part in the formation of, or to be an integral part of, important enzyme systems. It is generally accepted that all of them are required by all fungi and that fungi differ, not in their nutritional requirements, but in their synthetic powers. While some are completely dependent on an external source of a growth-substance (or substances), others are able to grow slowly in its absence but grow better if it is supplied. These are said to be 'partially deficient for' the particular growth-substance. Their synthetic ability is obviously insufficient to provide enough for optimum growth. Partial deficiency may take the form of an inability to sporulate without added growth-substance even when vegetative growth is fairly good, indicating that synthesis takes place too slowly to supply the needs of spore production before other metabolic changes become inhibiting.

Many examples of the beneficial effect of growth-substances on sporulation have been reported, but comparatively few attempts have been made to study the mechanism by which these act. In general, as with other types of nutrient, a favourable effect on mycelial growth is paralleled by an increase in asexual reproduction, but the conditions required for sexual reproduction may be more exacting or even quite different.

The effect of growth-substances on sporulation has been studied in more detail in the Pyrenomycetes than in any other group of fungi. Many members of this group are completely or partially deficient for thiamin or biotin or both, and some are also deficient for pyridoxine or adenine. The effects of these substances on the formation and maturation of perithecia and of viable ascospores has been studied in a number of species.

It has long been known that certain fungi produce fruit-bodies more readily in the presence of a colony of some other species. The first example described is that of *Melanospora pampeane* (Heald and Pool, 1908) which produces only scanty mycelial growth in pure culture but grows vigorously and produces numerous fertile perithecia in mixed culture with *Fusarium moniliforme*. Heald and Pool concluded that some chemical substance needed for the development of the perithecia of *Melanospora pampeane* is produced by *Fusarium moniliforme* and by certain other fungi, since extracts of the mycelium of these have a stimulating effect on growth and fruiting of *Melanospora pampeane* similar to that of the living colony. This conclusion was a notable one since, at the time of these experiments, little was known of the metabolic effect of growth-substances and vitamins and none of them had been purified and identified. From time to time other reports of the stimulating effect of one fungus on another have been published by various authors, and the effects have been variously attributed to reduction of the level of nutrients in the medium, modification of the H-ion concentration, the physical barrier to growth presented by a compact colony, and the production of hypothetical stimulating substances.

In 1936 Asthana and Hawker published the results of a

65

detailed study of the effects of a number of fungi on the production of perithecia by *Melanospora destruens*. They showed that, with any particular fungus as 'contaminant', the effects on *M. destruens* are stimulating (causing earlier and more numerous fruit-bodies), negative, or actually inhibiting, according to the dates of inoculation of the plates with the two fungi and their relative rates of growth. Not only can the effects of the 'contaminant' fungus on *Melanospora* be largely reproduced by filtrates of the medium in which pure cultures of the former have been grown, but this effect is limited to the ether-insoluble fraction of the filtrate and is actually increased by such fractionation. Asthana and Hawker concluded that the observed effects are due to the interaction of reduction of food supply by the 'contaminant' and the production by it of inhibitory staling substances (possibly organic acids, since these would have been removed by fractionation with ether) and of some substance or substances favouring fruiting. The stimulatory effect would be masked if the latter did not accumulate in the medium before an inhibitory level of staling substance was reached.

Hawker (1936) reported that an extract of lentils and some other natural extracts stimulates the production of perithecia by *M. destruens* in a similar manner. She suggested that 'the principle governing sporulation is that the fungus by its growth should decrease the concentration of nutrients below a certain point and increase that of the accessory substances above a certain point, before growth-inhibiting or staling factors have become too pronounced'. She further concluded that 'this favourable state of affairs is more likely to be reached if the medium be made up with a supply of accessory substances at the start'. Hawker compared the accessory substances to some others which had been reported as stimulating fungal growth, including a crude preparation of vitamin B_1. Later (Hawker, 1938, 1939a) a supply of pure thamin and biotin became available. By that time the strain of *M. destruens* in use had completely lost its ability to grow on a synthetic glucose-salts medium, on which it had previously grown and fruited sparsely. It still grew and fruited well if a small quantity of lentil extract or of media staled by certain fungi were added. This strain failed to grow when thiamin alone was added to the glucose-salts medium, grew fairly well but did not fruit when biotin alone was added and grew well and produced numerous fertile perithecia in the

66

presence of both thiamin and biotin (see Table 5). Thiamin also stimulates spore production in some other Pyrenomycetes and in various other fungi (Hawker, 1942).

TABLE 5. *The effect of biotin, thiamin (or its components), and inositol on growth and the formation of perithecia in* Sordaria (Melanospora) destruens. *(Data from Hawker, 1939a.)*

Growth-substances added to medium	Dry wt. mycelium (mg./100 ml. medium) after one week	Perithecial frequency
None	Negligible	None
Inositol (20 mg./100 ml. medium)	Negligible	None
Biotin (4 μg./100 ml. medium)	24	None
Thiamin (4 μg./100 ml. medium)	Negligible	None
Biotin + inositol	—	None
Thiamin + inositol	—	None
Biotin + thiamin	165	4·6
Biotin + thiamin + inositol	162	4·8
Biotin + thiazole component of thiamin	—	None
Biotin + pyrimidine component of thiamin	—	4·3
Biotin + thiazole + pyrimidine	—	4·7

Basal medium contained: glucose, 5·0 g.; KNO$_3$, 3·5 g.; KH$_2$PO$_4$, 1·75 g.; MgSO$_4$, 0·75 g.; distilled water, 1 litre. Perithecial frequency is the mean of counts of ten microscopic fields on each of three plates. A dash (—) means no data.

Since it was concluded that an initial supply of thiamin would make it more likely that conditions favourable for sporulation would be achieved, even where the fungus is able to synthesize a certain sub-optimal amount for itself, the effect of added thiamin on the initial concentration of glucose optimal for fruiting of *M. destruens* and some other fungi was studied (Hawker, 1942, 1944). Both thiamin and certain extracts (such as extracts of lentils or of mycelia) raised the level of initial concentration of glucose optimal for fruiting. It was shown that, with *M. destruens*, this is due to a more rapid removal of glucose from the medium with increasing concentration of thiamin, and that this can largely be accounted for by an increase in respiration rate which occurs in young cultures prior to fruiting. Since thiamin is an essential part of co-carboxylase, which controls the decarboxylation of pyruvic acid, the effect on respiration rate is readily understood. Other factors which increase the rate of respiration (such as the supply of sucrose instead of glucose, the presence of hexose phosphates, etc., see p. 56) also favour fruiting, and it may be concluded that such an increase provides energy, surplus to that required for vegetative growth, and which may be diverted to the synthesis of the

67

complex substances required in the development of the fruiting bodies.

A strain of *Sordaria fimicola* was shown by Hawker to resemble *Melanospora destruens* in requiring an external supply of both thiamin and biotin for maximum growth and perithecial formation. Another strain of this fungus was studied in greater detail by Lilly and Barnett (1947). This makes slow but sparse growth in a biotin-free medium but is entirely unable to produce either perithecia or any trace of sexual organs or fruit-body initials. In the presence of biotin, at concentrations of 0·1 to 6·4 μg. per litre, perithecia are present in all cultures by 42 days after inoculation, but are more than ten times as numerous at 6·4 μg. than at 0·1 μg. per litre. If, after 42 days in a biotin-free medium, the biotin-starved controls are supplied with biotin, numerous perithecia develop, showing that the mycelium has not lost the ability to produce fruit-bodies. The ratio of available food supply to biotin content of the medium is important. When this ratio is low vegetative growth is restricted and there is sufficient surplus biotin to permit reproduction, but when the ratio is high the available biotin is used up in profuse vegetative growth. Reduction in biotin concentration also reduces the percentage of asci which contain normal mature black ascospores.

Some other isolates examined by Lilly and Barnett behaved similarly, but one resembled the strain used by Hawker in requiring both biotin and thiamin for growth and reproduction. In a later paper (Barnett and Lilly, 1947) it was shown that the initial pH of the medium influences the rate and amount of growth at a particular level of biotin concentration. An initial pH of 3·8 to 4·0 is critical. Perithecia are never formed until the pH has been raised to 6·0 as a result of growth. The addition of thiamin to the medium prevents the subsequent drop in pH due to growth. Thus thiamin has a more favourable effect when added to media with a low initial pH. With such a low initial pH the time required for the development of perithecia decreases with increase in thiamin concentration. It was further shown that at very low pH values the synthesis of thiamin is inhibited and that this in turn inhibits growth and fruiting. The effect was further narrowed down to an inhibition of the pyrimidine component of thiamin. At higher initial pH values thiamin is readily synthesized and thus the effect of adding this vitamin is

68

less pronounced. It was suggested that the effect of thiamin in accelerating development of alkalinity in the medium was due to the transformation of pyruvic acid produced by the fungus. In support of this view it was shown that where the source of nitrogen was ammonium sulphate, leading to an accumulation of sulphuric acid in the medium, the addition of thiamin did not either neutralize the acidity or favour perithecial formation. The favourable effect of thiamin at low initial pH may thus be attributed to transformation of pyruvic acid as postulated by Barnett and Lilly. It might also be interpreted as a result of an increase in respiration rate, through the prevention of the accumulation of pyruvic acid, as with *M. destruens*.

Barnett and Lilly (1947*a*) also studied the effect of thiamin on *Ceratostomella fimbriata*. This fungus does not require an external supply of biotin but is partially deficient for thiamin. The addition of thiamin to a glucose medium increases both growth and the formation of perithecia. It has no effect on conidial production except as a result of increased growth. With increasing concentration of the medium the amount of thiamin needed is greater. If a thiamin-starved mycelium which has been grown on a medium containing 25 g. glucose and less than 2 μg. thiamin per litre is transferred to distilled water, no perithecia are developed, but if thiamin is added to the water, fertile perithecia develop in a few days. This shows that a well-nourished mycelium deficient in thiamin is unable to form perithecia but that, when it is deprived of all nutrients, except an adequate supply of thiamin, fruiting follows. Thiamin must, therefore, have a definite effect on the morphogenetic change from the vegetative to the reproductive phase.

Chaetomium convolutum, which is deficient for both thiamin and biotin, provides yet another example of the dependence of perithecial formation on the relative concentrations of nutrients and these growth-substances (Lilly and Barnett, 1949).

As a result of their studies Lilly and Barnett conclude that each unit weight of mycelium must contain a certain minimum concentration of both thiamin and biotin before sexual reproduction can take place. The results with *Melanospora destruens* further suggest that this essential internal concentration may be largely responsible for the increase in respiration which precedes the formation of perithecial initials.

Hawker (1936) showed that crude extracts, later shown

to contain biotin and thiamin, stimulated the formation of apothecia by *Ascobolus* spp. and *Saccobolus depauperatus*. Many Basidiomycetes are known to be deficient for thiamin. With both *Coprinus ephemerus* (Keyworth, 1942) and *Collybia velutipes* (Hawker, 1942) the amount of glucose optimal for fruit-body formation increases with increase in concentration of this vitamin. It is probable that biotin and thiamin play a similar part in the production of fruit-bodies by the Higher Ascomycetes and Basidiomycetes to that postulated with the Pyrenomycetes. *Schizophyllum commune* (Schopfer and Blumer, 1940) requires some unidentified growth-substance, in addition to thiamin, for the formation of normal fruit-bodies. Further detailed studies of the Higher Fungi are desirable.

The percentage conjugation among cells of the yeast *Zygosaccharomyces acidifaciens* is increased from 8–12% in pure cultures to 65–75% in cultures contaminated with *Aspergillus niger* (Nickerson and Thimann, 1941, 1943). This effect is also produced by filtrates of the culture medium in which the mould has grown, but not by adjustment of the hydrogen-ion concentration. It can also be reproduced by the addition of glutaric acid and riboflavin to the medium, but it is not known whether these two substances are the active factors in the *Aspergillus* medium.

Among the Mucorales, *Phycomyces blakesleeanus* has been studied by a number of workers. Schopfer (1931, 1931*a*) showed that this fungus is totally deficient for thiamin. Later it was shown (Schopfer, 1937; Robbins, 1939, 1939*a*; Robbins and Hamner, 1940) that the production of mature zygospores requires further growth-substances which are contained in certain samples of maltose and in such natural substances as cornmeal, oatmeal, potatoes and some samples of agar, and which were termed factor Z (Robbins, 1940, 1941). This was later thought to consist of two factors Z_1 and Z_2. Factor Z_1 was shown to be replaceable by the purine base guanine (Robbins and Kavanagh, 1942) and was later identified with hypoxanthine (Robbins and Kavanagh, 1942*a*; Robbins, 1943). Finally it was shown that factor Z_2 could be replaced by substances, or other factors, tending to prevent local accumulation of hydrogen ions. Robbins and his co-workers were able to show that hypoxanthine is not a specific factor for sexual reproduction, but that it influences vegetative growth also. The requirements for zygospore production are higher than those for mycelial growth.

More recently Hawker (unpublished work), in seeking an explanation of the inhibiting effects of low temperature on early stages in conjugation of *Mucor sexualis* and some other species of *Mucor* and *Zygorhynchus*, found that this can be partially overcome by the addition of any of the purines, guanine, hypoxanthine or adenine, to the medium. A mixture of one of these and thiamin has an even better effect. At favourable temperatures (20–25°C.) these substances induce earlier formation of zygospores. Adenine is the most effective. Growth rate is also slightly increased. One might suggest that the part played by these purine bases in the formation of nucleic acid, and consequently in protein synthesis, may account for their effect under conditions of insufficient synthesis or of reduced rate of diffusion and of protoplasmic streaming in the mycelium, such as are likely to be produced by low temperature. The addition of purines does not entirely overcome the inhibiting effects of low temperature.

Vitamins, particularly thiamin, are needed in external supply by a number of species of *Pythium* and *Phytophthora* (Robbins and Kavanagh, 1938, 1938*a*), but information as to the effect of these and of other unidentified substances on oospore production are contradictory. Raper (1952) showed that the beneficial effect of lentil extract on intensity of sexual reproduction in *Achlya ambisexualis* is due to the presence of *i*-inositol, in contrast to the results with *Melanospora destruens* on which this substance had no measurable effect.

An adequate supply of certain vitamins has also been shown to be necessary for *asexual* reproduction in a number of fungi, e.g. conidia of *Ceratostomella fimbriata* (Barnett and Lilly, 1947*a*), conidia of *Piricularia oryzae* (Leaver *et al.* 1947) and conidia of *Choanephora cucurbitarum* (Barnett and Lilly, 1950), but usually in amounts only slightly above those required for vegetative growth.

METABOLISM AND THE REPRODUCTIVE PHASE

It may be assumed, therefore, that while asexual reproduction does not involve any great increase in metabolic activity over that occurring during the phase of vegetative growth, sexual reproduction depends on a considerable intensification of such activity and thus requires a higher concentration of growth-substances and other activators. The problem of morphogenesis may best be attacked by careful comparisons of metabolic activity under conditions favourable and unfavourable to reproduction. Few such studies exist, but some examples may be considered which throw some light on the problem of factors controlling the change from vegetative to reproductive growth.

It has already been pointed out that asexual reproduction often takes place under conditions differing slightly or not at all from those favouring mycelial growth, but that the sexual or perfect stage usually depends on quite different conditions. The following remarks apply to the latter stage.

From the available evidence one can put forward two suggestions: (a) that anything tending to increase the actual rate of metabolism tends to encourage fruiting, and (b) that fruiting is the result of an interference with metabolism causing a change or 'shunt' in the paths of various syntheses in the direction of possible 'fruiting substances'.

At first sight these two ideas appear to conflict with one another, but further consideration shows that they are not necessarily mutually exclusive. It must be emphasized that fruiting is not a simple, or even a single process, but a sequence of events, each of which may be, and probably is, conditioned by a different set of chemical or other factors. Conditions favouring initiation and maturation are seldom identical, and it is not impossible that as the first fruit-bodies develop they may inhibit the development of others.

In support of the first hypothesis that fruiting is a result of an increased rate of metabolism, the work with *Melanospora destruens* already described may be cited. With this fungus any factor tending to increase the rate of respiration during the first four days of colony growth (such as ease of phosphorylation of sugars or high thiamin-sugar ratio in the medium) leads to the formation of earlier and more numerous fruit-bodies. In such cultures the bulk of the carbohydrate supply is used up before fruiting begins. Bretzloff (1954) showed that with high initial glucose, perithecial initials formed in cultures of *Sordaria fimicola* but afterwards aborted. An early increase in respiration rate has been correlated with increased and earlier fruiting in several other fungi, e.g. *Phycomyces blakesleeanus* (Hawker, 1944), while Burnett (1953) showed that respiration rate of *plus* and *minus* strains of heterothallic *Mucor* increased as the colonies approached one another, i.e. prior to conjugation.

72

The favourable effect of hypoxanthine and guanine, and of these or adenine, on zygospore production in *Phycomyces* and *Mucor* respectively, also suggests that an increased rate of metabolism (in this case of synthesis of nucleic acids or proteins) hastens and increases fruiting.

The effect of increase in temperature is certainly to increase the rate of metabolic processes. The effect of light on sporulation of some fungi might also be interpreted as a speeding-up of metabolic rate as a result of the supply of additional energy, but might equally well be interpreted as due to a change in the path of synthesis through the transformation of a light-sensitive precursor.

The well-established fact that dilution of the medium often leads to increased fruiting might be cited as evidence against the correlation of fruiting with an increase in metabolic rate. It must be emphasized, however, that dilution is often only effective with a previously well-nourished mycelium. It may well be that the release or translocation of metabolites from the older hyphae, which occurs under these circumstances, may itself lead to a temporary localized increase in metabolic rate sufficient to induce fruiting.

If an increase in the rate of metabolism is the sole internal factor leading to the initiation of fruiting, one would expect any intensification of a factor favourable to mycelial growth to bring about the change to the reproductive phase. This, however, is not the general rule, as is shown by the inhibition of sporulation of many fungi by concentrations of glucose, or other nutrients, considerably below that optimal for vegetative growth. The effect of the carbon-nitrogen ratio in determining the predominance of conidia or sclerotia in cultures of certain strains of *Botrytis* indicates that the two types of growth are induced by different metabolic sequences.

Hirsch (1954) and Westergaard and Hirsch (1954) trace a close correlation between the formation of pigments of a melanin type and the maturation of perithecia of *Neurospora crassa*. Both pigmentation and fruiting are favoured by factors inducing strong tyrosinase activity. The presence of melanin is not necessarily essential for fruiting and it is likely that the accumulation of the pigment is a result of the presence of surplus products of accelerated or diverted metabolism. It may have a biological value in the protection of spores against excess light. Its presence suggests, however, that the pathway of metabolism leading to reproduction differs from that leading to mycelial development only.

The most convincing work suggesting a 'shunt' or shift in the metabolic pathway as a cause of reproduction is that of Cantino (1951, 1952, 1953, 1954, 1956) with certain members of the Blastocladiales which produce two types of reproductive body; thin-walled zoosporangia and thick-walled resting bodies which produce swarmers only after a rest period. There is little doubt that the resting sporangia are of a sexual nature even when the swarmers have not been observed to fuse. Cantino showed that with *Blastocladiella emersonii* the zoospores will develop into either a thallus bearing thin-walled sporangia or one bearing the resting type according to the concentration of bicarbonate in the medium. Low concentrations induce the formation of zoosporangia, and higher concentrations cause the production of resting sporangia. After a certain critical stage of development is reached, changes in bicarbonate concentration no longer influence morphogenesis. Cantino considers that this is due to an accumulation of bicarbonate within the cell, accompanied by a decrease in permeability of the cell membranes, which prevents either loss of bicarbonate from the cell or further entry of this substance from outside. He interprets his results as being due to the effect of bicarbonate in decreasing oxidative decarboxylation of a-ketoglutarate and accelerating carbon-dioxide condensation with pyruvate, leading to the accumulation within the cell of substances intermediate between a-ketoglutarate and citrate. This accumulation of intermediates favours the development of thick chitinous walls and the accumulation of fatty reserve substances within the cell associated with a lowered rate of metabolic activity. Cantino and Hyatt (1953) describe a mutant strain which continues to grow vegetatively as long as the food supply is adequate. This strain has lost the enzymes, a-ketoglutarate oxidase and aconitase, which, in the normal strain, respond to the accumulation of metabolic carbon dioxide and bring about a limitation of growth and the production of resting bodies.

Nickerson (1954) shows how slight changes in the metabolic pathway are able to change the growth-form of certain fungi from a hyphal to a yeast-like phase. It is obvious that some such changes might lead to the initiation of the reproductive

74

phase. It is not impossible, however, that intensification of some metabolic process or processes might lead to the accumulation of excess products, which itself might induce a shunt in part of the metabolic sequence. In other words an unbalanced quantitative change could induce a qualitative change in metabolism. The problem can only be solved by detailed investigation of metabolism during the vegetative and reproductive phases of a number of fungi.

Meanwhile the effects of nutrition on sporulation may be summarized as follows. In general, high concentrations of readily available foodstuffs, and particularly of hexose sugars, inhibit sexual fruiting but may actually stimulate asexual sporulation. Complex organic substances may favour fruiting, which also requires higher concentration of certain minerals in the medium and of vitamins within the mycelium than are sufficient for mycelial growth.

CHAPTER 6

THE PHYSIOLOGY OF SEX

THE problem of sex in fungi is perhaps more complex than in any other group of organisms. In the Lower Fungi a definite fusion between clearly defined isogamous or heterogamous motile gametes or sexual branches usually takes place and results in the formation of a zygote. In the majority of the Higher Fungi, however, sex is not so clearly marked and a phase of secondary hyphal growth may intervene between the pairing of nuclei, which may or may not be derived from obviously differentiated male and female gametangia, and their ultimate fusion.

In the major groups of the Ascomycetes a series of stages in loss of sexuality may be traced. At the one extreme, nuclei migrate from a male branch, or antheridium, into a female cell (ascogonium) borne on a female branch, or archicarp, which consists of stalk cell (or cells), ascogonium and trichogyne or receptive cell. At the other end of the series all trace of sexual organs is lost. Between these extremes examples may be found lacking a trichogyne, or in which the antheridium is abortive, or in which only the female branch develops. In all these, however, a pairing of nuclei, whether male and female, both female or both vegetative, occurs and binucleate hyphae develop from the cell or cells in which this nuclear pairing has taken place. These ascogenous hyphae are usually larger and have denser contents than the vegetative hyphae. They may branch more freely. Their chief distinguishing character, however, is that they consist of binucleate cells. The two nuclei in a cell are known as a dicaryon and divide simultaneously so that one daughter nucleus from each goes into each daughter cell. This conjugate division continues until the initiation of the young asci, in which fusions of the paired dicaryotic nuclei take place. Attractive and repellent forces of an unknown nature keep the two nuclei of a dicaryotic pair together, prevent fusion until the development of the asci, and induce simultaneous divisions of the paired nuclei throughout the growth of the ascogenous hyphae.

76

Among Basidiomycetes, with the exception of the rusts, no trace of sexual branches or organs remains. Nevertheless in all the Basidiomycetes studied there is a clear alternation of a primary uninucleate, and usually sterile, phase, with a secondary binucleate phase which produces the fruit-bodies or, in the rusts and smuts, the teleutospores and brand spores. The problem of the origin and maintenance of the dicaryotic or binucleate phase in the Higher Fungi is discussed by Buller in his well-known *Researches on Fungi* (1909–50) and elsewhere (Buller, 1941).

Not only do the Higher Fungi exhibit the strange phenomenon of loss of sexual organs with the retention of nuclear fusions, followed by a reduction division during the formation of the so-called 'perfect' spore stage, but they may also lose the ability to produce these perfect spores altogether. The problem of this loss of fertility is one which has only been solved in a few examples.

Fungal hyphae of different strains of the same species show a marked tendency to anastomose. Anastomosis has been shown (Buller, 1941) to result in the migration and intermingling of nuclei of the two anastomosing hyphae. This leads to the well-known phenomenon of heterocaryosis in which the same hypha, and ultimately the same spore, may contain nuclei of different origins. Rearrangement of characters and variability of strains developing from single spores may thus result without the intervention of a sexual stage (Pontecorvo and Gemmell, 1944). Where anastomosis has taken place between mycelial and conidial forms of the same species the progeny resulting from single spore cultures may exhibit every gradation of sporulation (Hansen and Smith, 1932). This was termed the 'dual phenomenon' by Hansen (1938). Again nothing is known of the nature of the attraction which leads to anastomosis of hyphae and to subsequent migration of nuclei.

HETEROTHALLISM

The phenomenon of heterothallism was first observed by Blakeslee (1904) in the Mucorales. He showed that, while zygospores of certain species develop readily in cultures from single sporangiospores, those of other species do so only in a certain number of pairings of monospore cultures and that the zygospores then develop along the line of junction of the two colonies. Blakeslee termed the first group of species homothallic and the second heterothallic. He concluded that the heterothallic species exist in two strains, which he termed 'plus' and 'minus' and that zygospores only develop when both *plus* and *minus* strains are present. He also considered (1906) that the difference between the strains is a true sexual one, despite the fact that no constant morphological difference between the conjugating strains could be demonstrated. In support of this view he reported that imperfect hybridization between the heterothallic, isogamous *Mucor hiemalis* and the homothallic, heterogamous *Absidia spinosa* takes place only between the plus strain of the former and the smaller male gametangium of the latter or between the minus strain of the former with the large female gametangium of the latter. From this he deduced (Blakeslee, 1915) that the *plus* strain of a heterothallic species is female.

Since the publication of Blakeslee's work, heterothallism has been demonstrated in most other groups of fungi, including Oomycetes, Ascomycetes and Basidiomycetes. In the Higher Fungi it is often of a more complex nature than in the Mucorales. With some Ascomycetes either strain may produce both male and female branches, but does so only in the presence of the opposite or complementary strain as with *Ascobolus magnificus* (Dodge, 1920). With others fruit-body initials may form on a monosporous colony but continue their development only in the presence of a complementary strain, as with heterothallic species of *Neurospora* (Shear and Dodge, 1927). With yet other species, such as *Sclerotinia gladioli* (Drayton, 1932), receptive bodies of one strain develop into apothecia after 'fertilization' by microconidia of another strain. Thus the difference between heterothallic strains of Ascomycetes is not a sexual one. Similarly both strains of a heterothallic rust produce both spermagonia and aecidial initials, but the latter do not develop further unless 'diploidization' by the opposite strain has taken place (Craigie, 1927, 1927a, 1931; Buller, 1941).

The higher Basidiomycetes (Hymenomycetes and Gasteromycetes) include both homothallic and heterothallic species. The latter may be of the usual two-strain, or bipolar type or may be of a more complex four-strain or tetrapolar type. This has been shown to be due to two pairs of incompatibility

78

factors and may be further complicated by mutation from these, giving complete interfertility between the mutants, or between these and the parent strains. The existence of these multiple allelomorphs as a result of mutations explains the so-called 'geographical races', or interfertility between strains from different populations of the same species (Whitehouse, 1949, 1949a). Heterothallism in the smuts is of the two-strain type (Whitehouse, 1951). Claims for the existence of a more complex type have not been substantiated. The evolution of mating systems in fungi has recently been discussed by Burnett (1956a).

SEX HORMONES AND SPECIFIC 'FRUITING' SUBSTANCES

The problem of the nature of the difference between the complementary strains of a heterothallic fungus is a fascinating one and has yet to be solved. The recent acceleration of work on genetics of fungi has thrown considerable light on the problem and has shown conclusively that the difference between mating strains is a genetical one. As already stated, Blakeslee (1915) claimed that in the Mucorineae the difference was a true sexual one. While a case might be made out for such an interpretation in this group, it is obviously not tenable for the more complex types of heterothallism among the Higher Fungi. As early as 1928 Gwynne-Vaughan put forward the idea that the difference is likely to be a nutritional one. On this view each strain is able to synthesize some substance or to perform some step in a synthetic process essential for fruiting, of which the other is incapable, so that the two strains are complementary. Considerable evidence has been accumulated in support of Gwynne-Vaughan's hypothesis. Attempts to induce fruiting in single spore cultures of heterothallic strains by the application of extracts of the complementary strain, or of other fungi, have failed. One is thus forced to the view that we are dealing here, not with a general spore-promoting growth-substance, but with hormonal substances specific to the species concerned. Moreover, one must conclude that these are of such a labile nature that they cannot exist apart from the living fungus which synthesizes them.

Owing to the impetus to investigations given by Blakeslee's pioneer work and to the relatively simple life history of the Mucorineae, it is not surprising that this group received early

79

and extensive attention. Burgeff (1924) gave a detailed description of events leading up to conjugation between hyphae from the heterothallic colonies of *Mucor mucedo*. As the growing colonies of *plus* and *minus* strains of this fungus approach one another a 'zone of restraint' develops between them which is penetrated by only a few hyphae from each colony. The tips of these develop irregular swellings before they come into contact. Clearly some diffusible or volatile substance (or substances) produced by a particular strain influences the colony of opposite type, in advance of actual physical contact, so as to produce both inhibiting and stimulating effects. Burgeff then performed the classic experiment of separating the *plus* and *minus* strains by a permeable collodion membrane and showed that the initial restraint effect ('telemorphosis') and the mutual approach of hyphae of opposite strain ('zygotropism') still took place. He obtained similar results with *Mucor hiemalis*, *Rhizopus nigricans* and *Phycomyces nitens*, but no effect was observed when two colonies of the same strain of any of these were grown in a similar manner. Burgeff's work was the first example of the initiation and control of sexual reproduction by self-produced hormonal substances in fungi.

Results similar to those of Burgeff have been obtained for several members of the Mucorineae by a number of workers. Verkaik (1930) claimed that only the *minus* strain of *Mucor mucedo* produced a diffusible substance capable of attracting the other strain, but his results are based on rather poor experimental technique and are thus open to doubt. Kohler (1935) also studied *M. mucedo* and confirmed Burgeff's results that both strains produce diffusible stimulating substances. He was unable to determine the origin of these or their number or nature. Kehl (1937) also studied zygophore formation in *M. mucedo* and confirmed Burgeff's findings. Ronsdorf (1931) obtained similar results by the membrane technique with the heterothallic *Phycomyces blakesleeanus*. She was unable to reproduce the effect by the use of extracts of mycelia, but reported that the intensity of sexual response is increased by the presence of histamine. Krafczyk (1931, 1935) demonstrated a similar mutual attraction of hyphae of opposite strain and initiation of gametangial initials in *Pilobolus crystallinus*, in the absence of any physical contact between the strains. He concluded that initiation and subsequent development of the sexual reaction is

controlled by hormones up to the stage at which the gametangia are delimited.

Burnett (1953) produced additional evidence for the production of mutually stimulating substances by heterothallic
mycelia of *Mucor hiemalis* and *Phycomyces blakesleeanus*. He demonstrated that the consumption of oxygen increases significantly

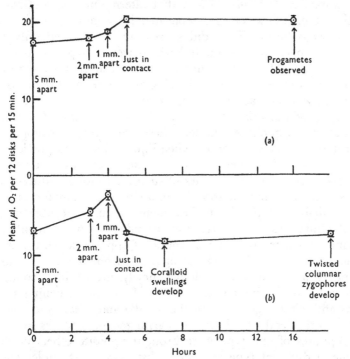

Fig. 4. Oxygen consumption by certain colonies of heterothallic Mucors. Mean
oxygen consumption of samples from the growing edges of compatible heterothallic
strains of (a) *Mucor hiemalis*, (b) *Phycomyces blakesleeanus*, prior to and immediately
after contact and the initiation of conjugation. $\odot = 2 \times$ standard deviation
(after Burnett, 1953).

just before two approaching colonies of opposite strain come
into contact (fig. 4). In view of the evidence that increased
rate of respiration may be a factor in the initiation of sexual
reproduction (p. 56) this is a most significant result. He was
also able to show (Burnett, 1953a) that, where opposite strains
of two different species were similarly tested, a similar, though
less pronounced, increase in oxygen consumption took place.

Burgeff (*loc. cit.*) refers to the conduction of the sexual stimulus through the air from one strain to another but gives no evidence for this. Raper (1952) was unable to find any published experimental evidence for such an air-borne stimulus. Recently Banbury (1954, 1955) grew strains of *Mucor mucedo* in a closed-ring system through which air was constantly recirculated, while allowing slow exchange with the outside atmosphere. He thus showed that alternating *plus* and *minus* colonies on blocks of agar separated by 3 mm. gaps did not form zygophores. They did so readily if separated only by sheets of permeable 'celloidin' which were not penetrated by the hyphae. Malt extract in which one strain had been grown was concentrated by partial freezing and subsequent removal of ice and was then added to assay cups placed near the edge of colonies of the opposite strain. Occasionally the *plus* strain produced zygophores under these conditions. Zygophore formation was also observed when liquid which had previously passed through a *plus* and then a *minus* culture reached a second *plus* culture in a series of such alternating cultures. It was concluded that the stimulus inducing zygophore formation is not air-borne but is due to the presence of diffusible substances.

Banbury next considered the second stage in conjugation, namely the attraction of zygophores of opposite strain towards one another. Small blocks of agar cut from the conjugation zone of *plus* and *minus* colonies newly grown into contact showed *plus* zygophores at one end of the block, conjugating *plus* and *minus* zygophores in the middle and *minus* zygophores at the other end. Banbury then arranged such blocks so that zygophores of one type were brought opposite others of the same or different type at various distances apart, under conditions precluding diffusion from one block to another. He claims that there was a mutual positive response over gaps of 2 mm. between zygophores of opposite strain and a mutual aversion between those of the same strain over gaps of 1 mm. and concludes that, in contrast to the stimulus controlling *zygophore formation*, that controlling direction of growth (zygotropism) can be transmitted through the air for a small distance. The stimulus could not pass through thin barriers of fused quartz, glass or aluminium foil unless these were perforated, in contradiction of Kohler's (1935) suggestion that the stimulus was due to electromagnetic radiation. Banbury suggests that two substances are concerned, one of which is produced by the *plus* strain and stimulates elongation of the zygophores of the *minus* strain while repressing those of the *plus* strain, and the other which is produced by the *minus* strain and has a similar effect on the *plus* one.

This work is the first attempt to distinguish between the stimuli causing the formation of zygophores and those inducing their subsequent elongation and directional growth. It is also the first experimental evidence for the volatile nature of any of these stimuli.

Banbury's work also supports Krafczyk's hypothesis that conjugation in the Mucorineae is controlled by hormones during the early stages.

Although it is thus established fairly conclusively that the mutual stimulation of opposite strains of the heterothallic Mucorineae is due to complementary hormones, the nature of these is obscure. Satina and Blakeslee (1925, 1926, 1926*a*) reported some chemical differences (including differences in the Manilov reaction, catalase content and ability to reduce tellurium salts and some other substances) between opposite strains. They compared these to male and female plants or plant organs, which have been reported to show similar chemical differences, and concluded that the *plus* strains of *Mucor* are female. Their work has never been successfully repeated and their conclusions are based on insufficient evidence to be generally accepted. Mating reactions of strains could, however, be predicted on the basis of their chemical characteristics. The strains showed no differences in internal H-ion concentration but there were possible differences in total acid production. Foster and Waksman (1939) claimed that, while the *plus* strain of *Rhizopus nigricans* produces abundant fumaric acid, the *minus* race produces none. It is unlikely, however, that fumaric acid or any other relatively stable substance is the hormone concerned since, with the exception of Banbury's inconclusive experiments, no attempt to reproduce the effect of one strain on another by the use of mycelial extracts has been successful. It is most likely that the substance or substances responsible are of a highly labile type and are destroyed, probably by oxidation, on separation from the living mycelium. In Banbury's partially successful experiments, the technique employed might be expected to overcome this difficulty to some extent.

A further experiment of Burgeff (1924) is of interest, although subsequent workers have been unable to confirm it. He studied species of Mucorineae parasitic on others and claimed that the *plus* strain of *Mucor* (*Parasitella*) *simplex* parasitizes only the *minus* strain of its host, *Absidia glauca*, while the *minus* strain of the parasite occurred only on the *plus* strain of the host. He interpreted this as attempts at hybridization, but the effects could better be interpreted as due to the nutritive requirements of the two strains of the parasite being differentially satisfied by the presumably chemically different strains of the host. Burgeff's results and conclusions are not accepted by Satina and Blakeslee (1926a).

Considerable attention has been given to the possibility of specific hormonal substances inducing the development of antheridia and oogonia in species of *Phytophthora*. Gadd (1925) reported that oogonia and oospores occurred when certain strains of *P. faberi* were mated but not with other combinations of strains, and concluded that this species is heterothallic. Further work with this species (Lester-Smith, 1927) suggested that the strains were weakly self-fertile but that sexual reproduction was stimulated in certain mixed cultures. The presence of the homothallic *P. parasitica* also caused an increase in spore production by single strains of *P. faberi*. A number of similar experiments with various strains and species of *Phytophthora* were reported by Ashby (1928, 1929, 1929a, 1929b). He finally came to the conclusion that the genus was entirely homothallic and that the effect of one strain or species upon another was to increase the intensity of sexual reproduction of normally weakly self-fertile strains by chemical stimulation. One of the difficulties of this type of work with *Phytophthora* is the abundance of the mycelium on any media suitable for sexual reproduction, and the consequent difficulty of tracing the origin of antheridial and oogonial branches. Narashiman (1930) attempted to reduce the number of hyphal branches concerned in reproduction by separating the strains by perforated mica sheets. By this method he claimed to have observed unions between hyphae of different strains of *P. arecae*. Leonian (1931) mated over 80 isolates of *P. omnivora* in various combinations and concluded that these fell into four distinct groups according to their sexual behaviour. Some strains were definitely homothallic, the majority were strictly heterothallic,

84

others were heterothallic but varied in the intensity of their response, and yet others were neutral, that is, they would mate with heterothallic strains of either type. Such a pattern of sexuality has been demonstrated for other groups of Oomycetes (Bishop, 1937, 1940; Bruyn 1935, 1937; Couch, 1926; Raper, 1936, 1939).

Kreutzer *et al.* (1940) described lines of oospores forming at the junction of strains of *P. capsici* isolated from cucumber fruits with those isolated from chilli stems. Barrett (1948) demonstrated that two isolates of *P. drechsleri* were male and female respectively, and Cohen (1950) obtained hybrid oospores between strains of *P. palmivora* and *P. cinnamomi*. Kouyeas (1953) reported that only 18% of a number of strains of *P. parasitica* produced oospores in single strain cultures, but that these were produced rapidly in suitably paired cultures. All strains were bisexual, however, but Kouyeas considers that sexuality is relative in this species. Stamps (1953), by a slide-culture technique, showed that hybridization can take place between various strains of *P. cinnamomi* and *P. cryptogea*. In these, *P. cinnamomi* provided the antheridium while the oogonium was derived from hyphae of *P. cryptogea*. Later, however, typical oospores of *P. cinnamomi*, differing morphologically from the hybrid ones, developed on the same plates, presumably as a result of chemical stimulation, due to the presence of the *cryptogea* strain, or produced by the hybrid oospores during their development. It is not clear in their various experiments whether the substances stimulating oospore production were specific sex hormones or were general metabolites of the growth-substance type, described in the previous chapter. Galloway (1936) induced the formation of oospores in single strains of *P. meadii* and *P. colocasiae* by the use of filtrates from paired cultures. Leonian (1935, 1936) described a growth-substance in pea extract and produced by certain green algae which he claimed to be a specific fruiting-substance of 'auxin' type, but he was unable to identify it. Zentmyer (1952) claimed that the water-soluble, non-protein content of an extract of avocado roots had a specific effect in stimulating sexual reproduction of *P. cinnamomi*. These last results suggest that some, at least, of the stimulating effect of some natural substances and of certain strains of *Phytophthora* on the fruiting of others is due to non-specific growth-substances.

85

The evidence for specific sex hormones in the Saprolegniales or water moulds is much more convincing. De Bary (1881) was the first to suggest the possibility that the initiation of antheridia of a species of *Saprolegnia*, and directional growth of these towards the oogonial initials, might be due to secretions from the latter. Kauffmann (1908) supported this view. Couch (1926) interpreted the production of antheridial hyphae and oogonial initials on male and female mycelia of *Dictyuchus monosporus* at points some distance from the zone of contact and the growth of antheridial hyphae towards the young oogonia, as evidence for hormonal control of these early stages in sexual reproduction. He also obtained some sexual interaction between the male thallus of *D. monosporus* and the homothallic *Thraustotheca primoachlya*. He was not able to obtain experimental evidence for the secretion of sex hormones.

Bishop (1937, 1940) obtained definite evidence for such a secretion of sex hormones by *Sapromyces reinschii*. He was able to induce the development of antheridial hyphae in male mycelia, growing on agar, by the addition of a cell-free filtrate of water in which female mycelia had grown. No similar response was shown by the addition to female cultures of filtrate in which male mycelia had grown. When male and female thalli were mated on agar plates, antheridial hyphae were frequently produced before contact with the female colony took place, and occasionally oogonial initials were similarly produced on the latter. Growth of the antheridial hyphae was always towards the oogonial initials. Bishop concluded that the production of antheridial branches was a direct response to diffusible substances produced by the female mycelia, and thought that the initiation of oogonia and the growth of the antheridial branches towards them were similarly under hormonal control.

A clear proof of the production of sex hormones by species of *Achlya* has been supplied by Raper in a series of papers describing ingenious and careful experiments (Raper, 1939a, 1940, 1940a, 1942 1942a, 1950, 1950a, 1951; Raper and Haagen-Smit, 1942). Raper worked with two dioecious species, *A. ambisexualis* and *A. bisexualis*, which bear their antheridia and oogonia on morphologically similar, but sexually distinct, thalli. The initiation and differentiation of sex organs takes place in five distinct stages: (i) production of fine branched

86

antheridial hyphae on the male plant; (ii) production of oogonial initials on the female; (iii) directional growth of the antheridial hyphae towards the oogonial initials, and the delimitation of the antheridia by the formation of cross-walls behind the tips after these tips have come into contact with the oogonial wall; (iv) delimitation of the oogonia by the formation of basal septa, and the differentiation of the protoplasm in the oogonia to give several spherical oospheres; and (v) the penetration of the oosphere wall by fine fertilization tubes produced by the antheridium and the migration of the antheridial nuclei through these to the oospheres. The regular and definite sequence of these events is itself evidence that the whole process of sexual reproduction is likely to be controlled by chemical substances or hormones.

Raper was able to demonstrate that sexual reaction took place before the mycelia came into contact, or when they were prevented from meeting. When two compatible thalli were mated in water they reacted vigorously along the line of contact, but a few antheridial hyphae and oogonial initials developed on the male and female thalli respectively at points remote from the zone of contact. If two such strains are inoculated at opposite sides of an agar plate, the antheridial hyphae may develop when the two colonies are as much as 18 mm. apart. Oogonial initials develop later but before the colonies meet. The stimulus is also perceived when the male and female thalli are separated by a semi-permeable membrane. The membrane prevents the completion of oospore production and the process is stopped at the delimitation of antheridia and oogonia. In the absence of oogonial initials the antheridia continue to elongate, even when in contact with the membrane, but with the formation of oogonial initials on the other side of the membrane they develop typical branched swellings wherever they are actually in contact with the membrane and these become differentiated into antheridia. Raper claims that this shows that both a chemical and a contact stimulus are necessary for the formation of antheridia.

The most convincing evidence for chemical control of sexual reproduction was obtained by Raper by a perfusion technique. A slowly flowing stream of water was passed successively over a series of microaquaria containing alternate male and female thalli. By this means it was shown that the initiation of the complex sexual reaction depends on the secretion of a substance or substances by the female thallus. This, which Raper denoted 'hormone A', induces production of antheridial hyphae on the male thallus. The production of oogonial initials is induced by a second hormone (B) produced by the male thallus only after its activation by hormone A. After the production of oogonial initials a third substance, hormone C, is produced by the female thallus which attracts the antheridial branches and these then delimit the antheridia on contact of the tips with the oogonium. The oogonium then becomes cut off by a basal septum as a result of the production of a fourth substance (hormone D) by the antheridia after delimitation. Raper considers it likely that the process of fertilization is also under hormonal control but was unable to devise any experiment to test this hypothesis.

Further work showed that hormone A is not a single substance, but that this initial stage in the reaction is controlled by at least four distinct substances. Of these hormones A and A^2 are produced by the female mycelium and, singly or in combination, are able to induce the formation of antheridial branches. These two substances are soluble in acetone and dioxane. Hormones A^1 and A^3 are secreted by the male thallus and are water-soluble and acetone-insoluble. They are unable to induce the formation of antheridial branches, but A^1 augments the activity of A and A^2, while A^3 depresses it. External factors, such as temperature, hydrogen-ion concentration, concentration and nature of nutrients in the medium, also influence the intensity of the initial sexual reaction. The difficulty of obtaining these substances in sufficient quantity for chemical analysis has so far prevented their identification, but certain saturated dicarboxylic acids induced the formation of a limited number of antheridial initials. Glutaric acid was the most effective and also increased the amount of hormone secreted by the female thallus. Malonic, pimelic and barbituric acids and hydantoin were also effective. Fig. 5 gives a diagrammatic scheme of the whole process of sexual reproduction in *Achlya*.

Recently (1955) Fischer and Werner have shown that a number of fungi, including the hyphae and antheridia of *Saprolegnia ferax* show positive chemotropism towards certain amino acids.

Raper also demonstrated that certain homothallic species of *Achlya* and species of the related genus *Thraustotheca* possess sexual mechanisms resembling that of the heterothallic *Achlya* spp. Interspecific and intergeneric mating of homothallic mycelia with male and female thalli of heterothallic strains

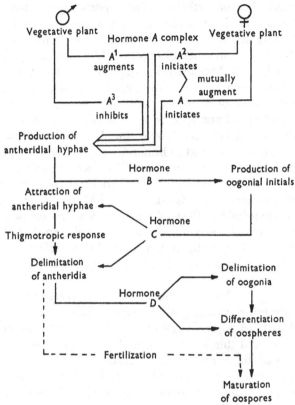

Fig. 5. The hormonal mechanism which co-ordinates the sexual interactions between male and female thalli of species of *Achlya*. For explanation see text, pp. 87–8 (after Raper, 1951, 1952).

yielded all degrees of reaction from complete indifference to complete compatibility, but most often the chain of reaction failed to continue beyond a point characteristic of the particular combination of strains. This suggests that the hormonal substances produced by different strains or species are similar but not identical.

The sexual anomalies described by Leonian (1931) for the genus *Phytophthora* (see p. 84), and by Humphrey (1892),

Maurizio (1899), Coker (1923) and others, in the water moulds, may be explicable in terms of the intensity of production of various parts of the sex hormonal system by different strains of the same species and by related species and genera.

The Higher Fungi are not so easily handled in this type of experiment, but nevertheless some reports suggest that sexual reproduction in these is also under hormonal control. There have been many instances of the stimulation of fruiting in one organism by the presence of another (Molliard, 1903; Sartory, 1912, 1916, 1918; McCormick, 1925; Wilson, 1927; and Dade, 1937). It is probable that with most of these the effect was due to a modification of the medium by reduction in food supply or by the synthesis of non-specific growth-substances as was shown by Asthana and Hawker (1936), for *Melanospora destruens* (see p. 66). Nickerson and Thimann (1943) showed that the apparently specific effect of *Aspergillus niger* on conjugations in *Zygosaccharomyces* could be reproduced by a mixture of ribo-flavin and glutaric acid. Goidanich and Mezetti (1948) showed that the stimulating effect of *Fusarium culmorum* on perithecial production by *Melanospora damnosa* was similar to that of biotin and thiamin. It is unlikely that any of these effects were due to specific sex hormones of the type described for species of *Achlya*.

Levi (1956) observed that cells of *Saccharomyces cerevisiae* frequently put out conjugation processes before they came into contact with an adjacent cell of opposite sex. On one occasion a single cell did this when separated by a collodion membrane from cells of the complementary strain.

Moreau and Moruzi (1931) claimed to have demonstrated the presence of sex hormones in two strains, M and N, of the pyrenomycetous *Neurospora sitophila*. In pure culture each of these produced only small sclerotia-like protoperithecia, but when inoculated into the two ends of a U-tube containing agar the N strain produced rather larger 'sclerotia' and the M strain produced fertile perithecia. This was interpreted as the result of diffusible substances passing through the agar. Dodge (1931) repeated this experiment and found that perithecia developed only when shrinking of the agar led to the presence of air spaces through which the hyphae of one strain had grown and inter-mingled with those of the other. Aronescu (1933, 1934) showed by an examination of the asci from perithecia forming at one end of such a U-tube culture that segregation of factors derived from each parent strain, in Mendelian ratio, had taken place, thus proving conclusively that the two cultures had intermingled, bringing compatible nuclei together.

Moreau and Moreau (1938) later produced new evidence in support of their claim of hormonal activity in *Neurospora*. Agar plates on which one strain was growing were heated for 10 minutes and the plates were inoculated with the complementary strain. If the temperature to which the plate had been subjected was as high as 110°C., no perithecia were produced by the second strain, but if it did not exceed 90–94°C., perithecia developed on the mycelium of the second strain. This they interpreted as due to the inactivation of the hormone at the higher temperature, but it might equally well have been due to failure to kill all the mycelium at the lower one. In the absence of genetical analyses of the ascospores the experiment is inconclusive. Lindegren (1934, 1936) reports that two strains which did not produce perithecia in pure or mixed culture did so when mated with a fertile mutant. This he considered to be due to the action of hormones produced by the fertile strain.

Derx (1926) described an unusual form of heterothallism in *Penicillium luteum*. Fertile cleistocarps (perithecia) developed only along the line of junction of appropriate mycelia derived from single ascospores, but the intensity of cleistocarp production varied according to the vigour of the mated strains. Thus a sexually vigorous monosporous mycelium produced cleistocarps which, however, did not develop ascospores. A sexually weak mycelium produced no fruit-bodies at all. The pairing of two vigorous mycelia led to the production of numerous fertile fruit-bodies, while only a few were produced when two feeble mycelia were paired. Mating of a vigorous with a feeble strain gave an intermediate number. This could well be interpreted as the result of differences in the amount of sex hormones produced by the various strains. However, later workers (Emmons, 1935; Raper and Fennell, 1952) have failed to demonstrate heterothallism in any of a large number of strains of *P. luteum*. In a discussion of Derx's results, Raper and Fennell point out that it is not unknown for homothallic cultures to produce heterothallic mutants (e.g. *Schizosaccharomyces pombe*; Leupold, 1950) or vice versa (*Ceratostomella fimbriata*; Olson, 1950), while in some homothallic species of *Neurospora*

(Dodge, 1927; Shear and Dodge, 1927), and a few other Pyreno-mycetes, dwarf spores occasionally arise which give rise to heterothallic mycelia. Derx may thus have worked with unusual mutant strains or with those developing from abnormal spores.

A more convincing example of incomplete heterothallism is that of *Glomerella cingulata*. Egerton (1914) showed that weakly self-fertile strains of this fungus produced perithecia in abun-dance along the line of junction of colonies and attributed this to the production of diffusible chemical substances by the mycelia. Driver and Wheeler (1955) mated self-fertile strains (producing clumps of perithecia) of this fungus with others which were self-sterile, or nearly so (producing only scattered perithecia), and obtained perithecia along the line of contact. These were of three types: (1) with spores giving rise only to cultures resembling the self-fertile parent, (2) those yielding cultures of both self-fertile and self-sterile strains, and (3) those giving self-sterile colonies only. The first type may be explained as the result of selfing of the self-fertile parent and the second as a cross fertilization between the fertile and sterile strains; but the origin of the third type is less obvious and Driver and Wheeler postulated the influence of hormonal substances to explain this. In experiments where the strains were separated by membranes the hyphae usually grew through these, and where this did not happen the effect of one strain on another was slight. Sterile filtrates of oatmeal media on which the fertile 'clumped' strain was growing induced better fruiting of weakly fertile strains. The ascospores from these almost invari-ably gave the 'scattered' type of colony, showing that the change induced by the filtrate was not genetic but hormonal, or at least nutritive. When fertile strains were grown on media unsuitable for fruiting, the filtrate was inactive. The greatest effect was on colonies 2–4 days old. Older colonies could not be stimulated in this way.

The evidence for hormonal stimulation of perithecial pro-ductions in *Glomerella* is thus suggestive, but it must be remem-bered that oatmeal is itself a particularly good medium for fruiting of many fungi. The growth of a colony on this medium might well make it even more favourable for fruiting when reinoculated with the same or a different fungus. This would not necessarily be due to the formation of specific hormonal substances by the first colony but might equally well be the result of reduction in one or more food substances or of change in acidity.

Backus (1939) showed that the trichogyne of *Neurospora sitophila* did not branch in single spore cultures, but that, if conidia or hyphal fragments of a complementary strain were present, the trichogynes produced one or more lateral branches which grew towards these. Similarly the trichogyne of *Bombardia lunatus* (Zickler, 1943) showed positive chemotropism towards masses of spermatia exuding from the spermagonia. The filtrate from a suspension of spermatia had a similar effect on the trichogynes, showing that this was due to a diffusible and relatively stable chemical substance.

In the Pyrenomycetes, such as heterothallic species of *Neurospora*, the evidence suggests that fruiting bodies are often initiated and develop to some extent on a single strain mycelium, but that development continues only in response to the stimulus of the presence of the complementary strain. Among Discomycetes, Dodge (1920) showed that antheridia and ascogonia of *Ascobolus magnificus* did not form in single-strain cultures but did so if two compatible strains were mated, suggesting that with this fungus, the earliest stages in fruiting are controlled by hormones. He further showed that the female cell or ascogonium did not continue its development unless the trichogyne remained in contact with the antheridium and he assumed that this phase was also regulated by a chemical mechanism.

Even less evidence exists for the production of sex hormones in the Basidiomycetes, but again the nature of the association of nuclei of opposite strain and the orderly sequence of events leading to basidiospore production suggests chemical control of reproduction. Buller (1941) discussed the problem of hyphal fusions and the mechanism controlling them but was unable to come to any precise conclusion owing to the essential difficulty of experimentation. Raper (1952) suggested that anastomosis might be controlled by a single species-specific labile substance secreted in minute quantities by every growing hyphal tip. He pointed out that experimental proof was rendered almost impossible owing to the fact that the responsible agents are effective only over a distance of 10 μ and must therefore be secreted in exceedingly minute quantities. Harder (1927) postulated hormonal control of clamp connexion formation in

Schizophyllum commune but produced no experimental evidence for this.

While no clear experimental evidence exists for the production of sex hormones in the Higher Basidiomycetes, several observers (Brunswik, 1924; Oort, 1930; Vandendries and Brodie, 1933; Vandendries, 1934; Brodie, 1935, 1936) have described mutual repulsion between certain strains of some species, which has been termed the 'barrage sexuel'. When two such colonies approach one another a gap is left between them and the aerial hyphae bend away from one another. Thin plates of glass, or of other non-permeable substances, placed in the medium but projecting only a little way above it, do not prevent the transmission of this effect which must, therefore, be due to some volatile substance or substances. Not all pairs of sexually incompatible mycelia show this mutual repulsion, which has been shown to occur only between mycelia containing certain nuclear combinations. Repulsion also occurs between dicaryotic and monocaryotic strains of the same species. It is thus clear that sexual strains of the Higher Basidiomycetes differ in their synthetic powers in some way not yet determined.

Thren (1937) attempted to explain the failure of the *plus* strain of *Ustilago nuda* to grow on potato agar and malt gelatine, on which the *minus* strain grew well, to lack of some substance necessary for growth which was presumably synthesized by the *minus* strain but not by the *plus* strain.

The phenomenon of 'relative heterothallism' discovered by Hemmons *et al.* (1953) may indicate hormonal attraction between compatible nuclei. Here when strains of *Aspergillus nidulans* were crossed the resulting cleistocarps showed more than the expected 50% of 'crossed asci', that is of asci containing spores of different nuclear type. This suggests a mutual attraction between nuclei of different types compared with those of the same type.

It is intrinsically likely that, as with the simpler lower fungi, so with the more complex higher ones, the whole process of sexual reproduction is co-ordinated by the production of a series of specific sex hormones. Definite evidence for this is slight, but such evidence as exists tends to support the hormone theory. That observations of the stimulating effect of one strain on another must be interpreted with caution, unless backed by precise experimental evidence of the type achieved by Raper with *Achlya*, is clear. Added emphasis to this cautionary note is given by the report by Das Gupta (1933) that the intermingling of two infertile strains of the imperfect fungus, *Cytosporina ludibunda*, led to the production of asexual pycnidia. Here, as with many examples of interstrain stimulation of true sexual reproduction, the cause may well have been the production by the two strains of separate essential growth-substances of a non-specific type (see p. 65) rather than of specific sex hormones.

THE DETERMINATION OF SEX

Coker (1923) claimed that the formation of antheridia in the Saprolegniales depends on the presence of phosphates in the medium to a greater extent than does the formation of oogonia. Studies of the separate nutritional requirements of male and female organs in other fungi are lacking.

The specific sex hormones are of almost unknown nature. Recently, however, attention has been given to the association of carotenoid pigments with one sex or the other in certain fungi. Carotenoids are of widespread occurrence in fungi and are often present in vegetative or asexual structures where they have been assumed to act as light-sensitive substances. In view of the possible function of carotenoids in reproduction of many plants and animals (Goodwin, 1950), an examination of the possible correlation of these substances with the reproductive processes in fungi is likely to be of interest. Carotenoids are frequently associated with spores and spore-bearing organs and are often lacking in vegetative mycelia of the same species.

Plus strains of *Mucor* have been said to contain a higher concentration of carotenoids than *minus* strains (Lendner, 1918; Satina and Blakeslee, 1926; Chodat and Schopfer, 1927). Schopfer (1943) claimed that the same is true for *Phycomyces blakesleeanus*, but this was contradicted by Garton *et al.* (1950, 1951), who worked with strains of which the *minus* form produced twice as much carotene as did the *plus* form. Barnett *et al.* (1956) demonstrated an increased production of carotene by *Choanephora cucurbitarum* when both *plus* and *minus* strains were present. Burnett (1956), however, has re-examined the evidence for a correlation between carotene and sex in *Mucor* and has concluded that the two are not correlated.[1]

[1] See also Carlile. *J. gen. Microbiol.* **14** (1956), 643–54.

More definite evidence of a constant association of carotenoids with one sex only has been obtained for species of the aquatic phycomycete *Allomyces* by Emerson and Fox (1940). The life history of species of this fungus of the sub-genus *Euallomyces* shows a regular alternation of an asexual (sporophytic) stage and a sexual (gametophytic) generation. The asexual plants bear thin-walled colourless zoosporangia and thick-walled resistant sporangia which have a dark melanic type of pigment. Spores from the former give rise only to asexual plants while those from the latter produce gametophytes. These bear relatively large female gametangia, which are colourless and produce large colourless female gametes, and smaller male gametangia which are orange in colour and give rise to small orange male gametes. Fusion of these in pairs produces motile zygotes from which the sporophyte develops, thus completing the life-cycle. The orange pigment of the male gametangia and gametes is contained in oil droplets and finally disappears on germination of the zygote. The pigment was shown to be mainly γ-carotene, with small quantities of other isomers, such as β-carotene, present in addition in some species. In the sub-genus *Cystogenes*, in which a sexual stage is lacking, *Allomyces cystogena* contained no detectable carotenoids while *A. moniliformis* contained γ-carotene.

Turian (1952) showed that carotenoid synthesis in *A. javanicum* could be inhibited by diphenylamine, but that even when the reduction in carotene content was 95% the male gametes were differentiated and liberated normally. Despite reduced motility these gametes fused with the female gametes and the resulting planozygotes developed normally. This work was confirmed by Turian and Haxo (1954), who showed that treatment with diphenylamine reduced the γ-carotene content but that at the same time measurable amounts of phytofluene and a pale yellow pigment of carotene type developed. It may be concluded either that the activity of the male gametes depends, if at all, on only a very low concentration of γ-carotene, or that other polyenes can function instead. The asexual plants not only lacked the pigmented carotenes, but also the colourless C_{40}-polyenes.

In the isogamous heterothallic *Blastocladiella variabilis* (Harder and Sörgel, 1938) the isogametes are borne on separate plants and only one strain produces the orange pigment.

In his studies of the life cycle and metabolism of *B. emersonii* Cantino (see p. 74) has made some interesting discoveries, and has put forward a novel interpretation of these (Cantino and Hyatt, 1953, 1953a; Cantino and Horenstein, 1954). Zoospores derived from resting sporangia[1] normally give rise to three types of 'plants', orange (*O*), ordinary colourless (*OC*) and later colourless (*LC*), distinguished from each other not only by colour but by rate of growth, size, motility and viability of swarmers produced and the potentialities of the resting sporangia produced from them. Normally only 2% of the plants are orange and these differ from the colourless ones in lacking α-ketoglutarate oxidase activity, in which they resemble orange plants of *Rhizophlyctis*. A mutant strain of *Blastocladiella emersonii* produces only orange plants. Neither the orange swarmers from the orange plants of the wild type nor those of the mutant show true fusion with the colourless ones from colourless plants. It was observed, however, that a succession of such orange mutant swarmers became for a time partially fused with a colourless one, and the possibility of cytoplasmic exchange, similar to that in *Paramecium*, was suggested.

It was further suggested that biosynthesis of pigment, formation of (*O*) plants and variability in viability of swarmers are controlled by a cytoplasmic factor which Cantino and Hyatt called the 'gamma' factor. Staining with Nadi reagent showed blue-black cytoplasmic particles in the orange swarmers which were absent from the colourless ones, and it was suggested that 'these particles stainable with the Nadi reagent may correspond to our cytoplasmic factor gamma'. According to Cantino and Horenstein (1954) 'maleness' in *Blastocladiella emersonii* is 'determined not genotypically but, rather (i), by the random distribution of cytoplasmic particles during formation of swarmers in individual plants, and (ii), by the transfer of such particles via the temporary cytoplasmic bridge which is found between orange and colourless swarmers before they germinate to produce new plants'.

Hatch (1935) demonstrated a differential distribution of mitochondria in gametangia of *Allomyces*. These accumulate in the hyphal tips and are therefore most numerous in the terminal female gametangium. Cantino and Hyatt suggest that

---

[1] Emerson (1950) considers it likely that resting sporangia of all Blastocladiales resemble those of *Allomyces* and that meiosis occurs during their germination.

mitochondria might be the vehicle of their 'gamma' factor, the quantities of which in a swarmer controls its activity and pigmentation.

Cantino's work with *Blastocladiella* and that of various investigators with *Allomyces* suggest that the presence of carotene is closely linked with the determination of sex in motile gametes. The correlation between the presence of visible carotenoid pigments and sexual activity is not perfect, however, and does not constitute a proof of a causal relationship between these two phenomena. Moreover, the parallelism between the presence of yellow, orange or red pigments and spore production is seen in only comparatively few fungi. Thus, while the study of the relationship between carotenoid substances and reproduction is of great interest, it is unlikely that it will lead to the elucidation of the mechanism of either reproduction in general or of sex determination in fungi. As with Hirsch and Westergaard's studies of the correlation between melanin production and fruiting in *Neurospora*, it seems likely that pigments may represent accumulations of the products of metabolism and that they tend to be produced in large quantities when an enhanced rate of metabolism occurs, as in the phase preceding sporulation.

LOSS OF SEXUALITY

It is clear that many fungi have no sexual stage or true sexual nuclear fusion in their life history and many others develop parthenogenetically. These must have developed from types with a normal sexual stage. No evidence exists as to how and when this partial or complete loss of sexuality occurred.

In culture, however, it is common for fungi to show a gradual loss in intensity of sexual reproduction over a number of transplants. This may show itself in a reduction in the number of fruit-bodies formed or in the number of viable spores per fruit-body. In a number of fully investigated examples, such an apparently gradual loss of sexuality has been shown to be due to the unconscious selection of sterile, or relatively sterile, mutants during sub-culturing.

Mohendra and Mitra (1930) showed that, with repeated sub-culturing, a strain of *Sphaeropsis malorum* changed from a dark-coloured colony with abundant pycnidia, containing viable spores, to a white one which was almost sterile. When spores from a 'fertile' type of pycnidium were plated out they gave rise to both black and white colonies. Those giving rise to the sterile white colonies germinated more quickly than those producing black colonies and hence with mass inocula of spores the white colonies had a greater chance of developing.

Similarly, loss of the ability to produce perithecia by *Melanospora destruens* (Hawker, 1947) was shown to be due to the more rapid rate of growth at room temperature of sterile mutants than of the parent type of fertile strain. By incubation at 37° C. for a few days after sub-culturing the position was reversed, since the fertile strains showed the more rapid growth at this temperature.

Thus, in these examples, it is unlikely that loss of sexuality was a slow, continuous process and it probably took place as the result of a series of mutations.

In a preliminary account of experiments with *Aspergillus nidulans*, Jinks (1954) described the development of asexual and predominantly sexual strains respectively as the result of repeated transfer by conidia only or by ascospores. He interpreted this as due to changes in the cytoplasm, and thus assumed that cytoplasmic inheritance occurs in this fungus. His results are in marked contrast to those of Mohendra (1928), who obtained no visible differences in strains of a number of fungi through the use of spores or mycelium as inocula over a number of sub-culturings. It is clear that the whole question of loss of fertility in culture needs critical examination.

It is to be hoped that the study of the various aspects of sex in fungi will be extended to include more species.

CHAPTER 7

REPRODUCTION IN THE
NATURAL HABITAT

MOST of the results described in the previous chapters were derived from the study of fungi in pure culture. A pure culture of a fungus is seldom or never found outside the laboratory. Nevertheless the results obtained by the study of pure cultures enable us to interpret to some extent the relation between sporulation and environment in the infinitely more complex conditions obtaining in the natural habitat.

SUMMARY OF CONCLUSIONS FROM PURE
CULTURE STUDIES

A young fungal colony does not usually produce spores until after a preliminary vegetative phase even under the most favourable environmental conditions (p. 2). Reproduction takes place only when both internal and external conditions are favourable. Little is known of the nature of the changes within the cells which enable the reproductive phase to begin. The duration of this preliminary vegetative phase in the life of a colony is likely to be of great importance in the natural habitat when competition from other organisms and the risk of desiccation make the rapid production of spores essential for survival of species with a delicate and unprotected mycelium.

Chapter 4 (p. 24) shows that temperature, water content of the substrate, relative humidity of the atmosphere, aeration and hydrogen-ion concentration of the substrate, all have a profound influence on the rate and volume of spore production and sometimes on the type and morphology of the spores and fruit-bodies. Sporulation frequently takes place only over narrower ranges of these factors than those sufficing for mycelial growth. Again, some fungi are induced to sporulate by slight injury, such as the presence of toxins in very low concentration, sub-lethal doses of ultra-violet light or mechanical injury

(p. 37) or by the thigmatropic stimulus of contact with a hard substance (p. 38). Light influences sporulation in a variety of ways (p. 38), while the gravitational stimulus controls the growth of fruit-bodies of many Higher Fungi and thus plays a part in the mechanism of spore discharge (p. 45). All these factors are likely to be as important in the natural habitat as in pure culture.

It was concluded (Chapter 5) that nutrition is the most important single factor controlling reproduction. In the natural habitat, food supply is often a limiting factor. Competition among species and individuals for the available food materials has been shown by many observers to control the nature of the microflora of a particular habitat and it has undoubtedly also a controlling influence on sporulation. Pure culture work has shown that the concentration of the medium or of certain constituents of it is of great importance. While sporulation, and particularly the production of large fruit-bodies, requires a larger total *amount* of food than is sufficient to maintain a minimum of mycelial growth, it is often inhibited at much lower *concentrations* than those which inhibit growth. The *nature* of the food supply is also important and the requirements of reproduction often differ from those of vegetative growth. Complex carbohydrates, such as starch and sucrose, are often tolerated in higher concentrations than are their hydrolysis products, the hexose sugars. This beneficial effect of polysaccharides on sporulation is correlated with the rate of breakdown and the ease of phosphorylation of them by the fungus (p. 55).

The effect of the carbon-nitrogen ratio has been studied in only a few species (p. 57) and may be of importance in the natural habitat. The effect of the nature of the nitrogen source varies with different species and no generalization can be made (p. 61).

In nearly all the examples which have been critically examined, phosphorus, potassium, magnesium, sulphur and many 'trace elements', such as iron, copper or molybdenum, are required in greater quantity for spore production than for hyphal growth. In at least one example calcium has been shown to have a beneficial effect on fruiting (p. 62). Shortage or excess of any one of these elements may limit sporulation under natural conditions.

The importance of vitamins for sporulation is now well

established (p. 64). Shortage of these in the natural habitat may well be a limiting factor in spore production.

Heterothallism (p. 77) is so widespread among fungi that many mycelia may remain sterile under natural conditions through failure to meet a mycelium of compatible mating type. The advantage of the large number of spores produced by fungi is obvious since, although only a small proportion of these are able to establish themselves in new surroundings, those that succeed must have a greater chance of encountering the complementary strain than would be possible if fewer spores were produced.

THE NATURAL HABITAT

(a) *Production of fruit-bodies by the larger fungi.* It is generally accepted that in Britain the large fleshy fruit-bodies of the agarics and of some Higher Ascomycetes develop mainly in late summer and early autumn. Wilkins and his co-workers (Wilkins and Patrick, 1940; Wilkins and Harris, 1946) studied the effect of weather on the incidence of these fruit-bodies and considered that the size of this peak production is controlled by the weather conditions obtaining during the previous months. A hot summer followed by a warm, wet autumn usually gives a maximum crop of fruit-bodies. The effect of rainfall and temperature on the fungus flora of a beechwood and a pinewood was studied. Wilkins and Harris concluded that to give an abundant crop of fruit-bodies, temperature must be above a certain level and rainfall must be within a certain range, and that favourable temperature and rainfall must operate over a certain period which varies for different species. Some evidence was obtained that species which regularly produce their fruit-bodies during the summer months have a slightly higher temperature optimum than those normally occurring in the autumn.

The present writer has regularly collected the larger fungi in a number of different woods in the Bristol area during the years 1945–55. The incidence of these has agreed with Wilkins's findings. In addition very wet summers, even when the temperature has been fairly high, have been followed by poor crops of fruit-bodies in the autumn, suggesting that aeration of the soil is an important factor, since fruiting does not occur when the soil is waterlogged. Recent laboratory experiments under controlled conditions (see p. 35) confirm this interpretation. The best crops in the Bristol area during this period were in 1948, when a warm, sunny period in late September and October followed a warm but wet summer, and in 1953 when a mild autumn followed a summer with considerable warm periods together with rain. The cold, wet summer of 1954 was followed by a very poor autumn crop. The dry, hot summer of 1949 was followed by an equally poor autumn crop and after the abnormally hot, dry summer of 1955 very few fruit-bodies developed and their production was soon checked by early frost. After the two peak years of 1948 and 1953 the following years were poor which, to some extent, supports the view held by some mycologists that the mycelium becomes exhausted by excessive fruit-body production and may need a long period to recover. It has been shown (p. 49) that fruiting usually takes place only on a well-nourished mycelium. However, in both the years 1949 and 1954 other conditions (i.e. drought and low temperature with excessive rainfall, respectively) could account for the poor crop. It seems unlikely that the mycelium, whether of mycorrhizal or saprophytic fungi, could not replenish its nutrients rapidly once favourable conditions recurred. It is, however, difficult to explain why some species fail to occur in what is, in general, a good year.

In a study of hypogeous fungi, Hawker (1954) recorded data of weather and soil conditions and of the plant species present, for over 1000 collections of subterranean fruit-bodies including representatives of the Tuberales, Elaphomycetaceae and Gasteromycetes. These records showed the importance of temperature and rainfall and the effect of aeration and hydrogen-ion content of the soil. Different species were favoured by different combinations of these factors. The effect of contact stimuli was shown by the frequency with which fruit-bodies of some species developed in contact with hard objects, such as the edge of a path, or against a hard pan of clay underlying the woodland litter.

With these hypogeous fungi the effect of environment depends to some extent on the different lengths of time taken for the fruit-bodies of various species to mature. Species of the Tuberales, including the true truffles, are initiated early in the year, as soon as conditions of moisture and soil temperature are both favourable, and mature in late summer or autumn.

103

Thus if environmental factors are unfavourable in the early part of the year they may prevent the initiation of fruit-bodies.

Unfavourable factors (such as drought) may delay development and thus reduce the final size of the fruit-bodies or, in extreme cases, may cause them to abort. The fruit-bodies of the Tuberales, however, are relatively tough in texture and usually have a well-developed peridium so that they are not so easily damaged by drought as are the more delicate fruit-bodies of most of the hypogeous Gasteromycetes. The latter may be initiated at any time in the year when weather conditions are favourable. Mature fruit-bodies of some (such as *Hymenogaster luteus, H. olivaceus* or *H. tener*) are usually to be found two to three weeks after the termination of a dry or cold period. They do not survive long in the soil, however, and readily abort at any stage through drought or extreme cold.

The Elaphomycetaceae are tough-skinned like the Tuberales and mature fruit-bodies are resistant to both drought and frost. Unlike the Tuberales they can be initiated at any time of the year when conditions are favourable. Consequently mature fruit-bodies of these fungi can usually be found at all times of the year except in very prolonged dry periods, but young ones are only found when the soil is moist and not too cold. Large numbers are usually present in suitable habitats shortly after a thaw following prolonged frost or with the return of damp conditions after a drought period. They are the commonest hypogeous species, and certainly owe their success in part at least to their ability to survive periods of drought and cold through their thick peridium, which is itself surrounded by a further protective layer, or crust, of soil particles interwoven and tightly bound by hyphae, and to their ability to initiate new fruit-bodies at any time of the year when conditions are suitable.

(*b*) *Plant pathogens.* The problems of plant pathology, too, may sometimes be interpreted by knowledge of the factors influencing sporulation. Byrde (1953) has applied such knowledge to the interpretation of the relationship between environmental factors and the production of conidia by *Monilia fructigena* (brown rot) on mummified fruits. Recent partially successful attempts to forecast epidemics of *Phytophthora infestans* (potato blight) (Beaumont, 1947; Large, 1953;

Hirst, 1955) are also based on the knowledge of the effects of weather conditions on the production of sporangia by this fungus. Many more examples could be given. Control of sporulation is often also control of the disease organism, as with the prevention of spread of the tomato mould (*Cladosporium fulvum*) by good ventilation of the houses and consequent low relative humidity.

Abnormal weather conditions often produce unusual spore stages in plant disease organisms. Thus in hot years cleistocarps of the oak mildew (*Microsphaera quercina*) may develop in England, where this fungus is usually entirely conidial (Robertson and Macfarlane, 1946). Similarly, the strain of antirrhinum rust (*Puccinia antirrhini*) which spread across England in the early 1930's was for long thought to be incapable of producing teleutospores and to maintain itself solely through uredospores. The writer observed teleutospores on diseased plants brought into a warm greenhouse, and in the hot summer of 1955 these spores were developed freely out of doors.

It has already been pointed out (p. 50) that changes in food supply may explain changes in spore type in plant disease fungi, many of which produce asexual spores in quantity during the summer and produce the perfect stage only in autumn or during the winter. The production of fruit-bodies of the wood-destroying polypores only after the destruction of the tree, or of most of its wood, is well known and is almost certainly correlated with changes in the nature of and reduction of the food supply.

Inhibition of reproduction, whether by spores or sclerotia, of a plant parasite may lead to its eradication by preventing either spread or survival or both.

The importance of the study of the physiology of reproduction to the ecologist and the plant pathologist is thus clear. It is of almost equal importance to the industrial mycologist who is seeking to prevent the spoiling of stored products through fungal attack. The fungi with which he is concerned are almost all of the type which produce quantities of air-borne conidia, and eradication of these spores would destroy the foci of infection. It is clear that the physiology of reproduction in fungi is of the greatest practical importance and of great fundamental interest in itself. A wide field for careful study remains. Future investigators would do well to concentrate on (1) the conditions within the cell which induce the change from the vegetative to the reproductive phases, (2) the effect of environment on particular stages in reproduction rather than on the process as a whole, and (3) the interaction of factors controlling sporulation in the complex natural habitat.

REFERENCES

AINSWORTH, G. C. (1952). *Medical Mycology. An Introduction to its Problems*, 273 pp., London.

ALLINGTON, W. B. (1936). Sclerotial formation in *Rhizoctonia solani* as affected by nutritional and other factors. *Phytopathology*, 26, 831–44.

ARENS, K. (1929). Untersuchungen über *Pseudoperonospora humuli* (Miyabe u-Takah) den Erreger der neuen Hopfenkrankheit. *Phytopath. Z.* 1, 169–93.

ARONESCU, A. (1933). Further studies in *Neurospora sitophila*. *Mycologia*, 25, 43–54.

ARONESCU, A. (1934). Further tests for hormones in *Neurospora*. *Mycologia*, 26, 244–53.

ASCHAN, KARIN (1954). The production of fruit-bodies in *Collybia velutipes*. I. Influence of different culture conditions. *Physiol. Plant.* 7, 571–91.

ASHBY, S. F. (1928). The oospores of *Phytophthora nicotinae* Br. de Haan, with notes on the taxonomy of *Phytophthora parasitica* Dastur. *Trans. Brit. mycol. Soc.* 13, 86–95.

ASHBY, S. F. (1929). Strains and taxomony of *Phytophthora palmivora* Butler (*P. Faberi* Maubl.). *Trans. Brit. mycol. Soc.* 14, 18–38.

ASHBY, S. F. (1929a). Further note on the production of sexual organs in paired cultures of species and strains of *Phytophthora*. *Trans. Brit. mycol. Soc.* 14, 254–60.

ASHBY, S. F. (1929b). The production of sexual organs in pure cultures of *Phytophthora cinnamomi* Rands. and *Blepharospora cambivora* Petri. *Trans. Brit. mycol. Soc.* 14, 254–63.

ASTHANA, R. P. and HAWKER, L. E. (1936). The influence of certain fungi on the sporulation of *Melanospora destruens* Shear, and of some other Ascomycetes. *Ann. Bot., Lond.*, 50, 325–44.

BACHMANN, J. (1895). Einfluss der äusseren Bedingungen auf die Sporenbildung von *Thamnidium elegans. Bot. Ztg.* 53, 107–30.

BACKUS, M. P. (1939). The mechanics of conidial fertilization in *Neurospora sitophila. Bull. Torrey bot. Cl.* 66, 63–76.

BAKER, R. E. D. (1931). Observations on the conditions for spore formation in *Sporodinia grandis* Link. *New Phytol.* 30, 303–16.

BANBURY, G. H. (1954). Processes controlling zygophore formation and zygotropism in *Mucor mucedo* Brefeld. *Nature, Lond.*, 173, 499–50.

BANBURY, G. H. (1955). Physiological studies in the Mucorales. III. The zygotropism of zygophores of *Mucor mucedo* Brefeld. *J. exp. Bot.* 6, 235–44.

BANNERJEE, S. N. and BAKSHI, K. (1944). On the production of true pilei of *Polyporus brumalis* (Pers.) Fr. in artificial culture. *Curr. Sci.* 13, 102–4.

BARNETT, H. L. and LILLY, V. G. (1947). The effects of biotin upon the formation and development of perithecia, asci and ascospores by *Sordaria fimicola* Ces. and de Not. *Amer. J. Bot.* 34, 196–204.

BARNETT, H. L. and LILLY, V. G. (1947a). The relation of thiamin to the production of perithecia by *Ceratostomella fimbriata*. *Mycologia*, **39**, 699–708.

BARNETT, H. L. and LILLY, V. G. (1950). Nutritional and environmental factors influencing asexual sporulation of *Choanephora cucurbitarum* in culture. *Phytopathology*, **40**, 80–9.

BARNETT, H. L. and LILLY, V. G. (1955). The effects of humidity, temperature and carbon dioxide on the sporulation of *Choanephora cucurbitarum*. *Mycologia*, **47**, 26–9.

BARNETT, H. L., LILLY, V. G. and KRAUSE, R. F. (1956). Increased production of carotene by + and − cultures of *Choanephora cucurbitarum*. *Science*, **123**, 141.

BARRETT, J. T. (1948). Induced oospore formation in the genus *Phytophthora* (Abstract). *Phytopathology*, **38**, 2.

BASU, S. N. (1951). Significance of calcium in the fruiting of *Chaetomium* species, particularly *Chaetomium globosum*. *J. gen. Microbiol.* **5**, 231–8.

BEAUMONT, A. (1947). The dependence on the weather of the dates of outbreak of potato blight epidemics. *Trans. Brit. mycol. Soc.* **31**, 82–91.

BILLE-HANSEN, E. (1953). Fructification of a coprophilous *Coprinus* on synthetic medium. *Physiol. Plant.* **6**, 523–8.

BILLE-HANSEN, E. (1953a). Fructification of three coprophilous species of *Coprinus* using glucose, sucrose and maltose as carbon source. *Sætryk. af Bot. Tidskr.* **50**, 81–5.

BISBY, G. R. (1925). Zonation in cultures of *Fusarium discolor sulphureum*. *Mycologia*, **17**, 89–97.

BISHOP, H. (1937). Sexuality in *Sapromyces reinschii*. Thesis Harvard University, Cambridge, Mass.

BISHOP, H. (1940). A study of sexuality in *Sapromyces reinschii*. *Mycologia*, **32**, 505–29.

BLAAUW, A. H. (1914, 1915, 1919). Licht und Wachstum. I. *Z. Bot.* **6**, 641–703. II. *Z. Bot.* **7**, 465–532. III. *Meded. LandbHoogesch., Daaraan Verbouden Inst.* **15**, 89–204.

BLAKESLEE, A. F. (1904). Sexual reproduction in the Mucorineae. *Proc. Amer. Acad. Arts Sci.* **40**, 205–319.

BLAKESLEE, A. F. (1906). Differentiation of sex in thallus, gametophyte and sporophyte. *Bot. Gaz.* **42**, 161–78.

BLAKESLEE, A. F. (1915). Sexual reaction between hermaphroditic and dioecious Mucors. *Biol. Bull., Woods Hole*, **27**, 87–102.

BONNER, J. T., KANE, K. K. and LEVEY, R. H. (1956). Studies on the mechanics of growth in the common mushroom, *Agaricus campestris*. *Mycologia*, **48**, 13–19.

BOWMAN, D. H. (1946). Sporidial fusion in *Ustilago maydis*. *J. agric. Res.* **72**, 233–43.

BREFELD, O. (1877). *Botanische Untersuchungen über Schimmelpilze*, Heft III, pp. 1–266. Leipzig: Verlag von Arthur Felix.

BRETZLOFF, C. W. JR. (1954). The growth and fruiting of *Sordaria fimicola*. *Amer. J. Bot.* **41**, 58–67.

BRIAN, P. W. and HEMMING, P. J. (1950). Some nutritional conditions affecting spore production by *Trichoderma viride* Pers. ex Fries. *Trans. Brit. mycol. Soc.* **33**, 132–41.

BRIGHT, I. B., DIXON, P. A. and WHYMPER, J. W. T. (1949). Effect of ethyl alcohol and carbon dioxide on the sporulation of baker's yeast. *Nature, Lond.*, **164**, 544.

BRODIE, H. J. (1935). The occurrence in nature of mutual aversion between mycelia of Hymenomycetous fungi. *Can. J. Res. C.* **13**, 187–9.

BRODIE, H. J. (1936). The barrage phenomenon in *Lenzites betulina. Genetics*, **18**, 61–73.

BROWN, W. (1923). Experiments on the growth of fungi on culture media. *Ann. Bot., Lond.*, **37**, 105–29.

BROWN, W. (1925). Studies in the genus *Fusarium*. II. An analysis of factors which determine the growth form of various strains. *Ann. Bot., Lond.*, **39**, 373–408.

BROWN, W. (1935). On the *Botrytis* disease of lettuce, with special reference to its control. *J. Pomol.* **13**, 247–59.

BROWN, W. and HORNE, A. S. (1926). Studies in the genus *Fusarium*. III. An analysis of factors which determine certain microscopic features of *Fusarium* strains. *Ann. Bot., Lond.*, **40**, 203–21.

BRUNSWIK, H. (1924). Untersuchungen über die Geschlechts- und Kernverhältnisse der Hymenomycetengattung *Coprinus. Bot. Abh. herausgegeben von Dr. K. Goebel*, **1**, 1–152.

BRUYN, H. L. G. DE (1935). Heterothallism in *Peronospora parasitica. Phytopathology*, **25**, 8.

BRUYN, H. L. G. DE (1937). Heterothallism in *Peronospora parasitica. Genetics*, **19**, 553–8.

BULLER, A. H. R. (1909–50). *Researches on Fungi*, vols. I–VII, London.

BULLER, A. H. R. (1941). The diploid cell and the diploidization process in plants and animals, with special reference to the higher fungi. *Bot. Rev.* **7**, 335–431.

BÜNNING, E. (1937). Phototropismus und Carotinoide. II. Das Carotin der Reiz-aufnahmezonin von *Pilobolus, Phycomyces* und *Avena. Planta*, **27**, 148–58.

BURGEFF, H. (1924). Untersuchungen über Sexualität und Parasitismus bei Mucorineen, I. *Bot. Abh.* **4**, 5–155.

BURNETT, J. H. (1953). Oxygen consumption during sexual reproduction of some Mucoraceae. *New Phytol.* **52**, 58–64.

BURNETT, J. H. (1953 a). Oxygen consumption of mixture of heterothallic and homothallic species in relation to 'imperfect hybridization' in the Mucoraceae. *New Phytol.* **52**, 86–8.

BURNETT, J. H. (1956). Carotene and sexuality in Mucoraceae, especially in *Phycomyces blakesleeanus. New Phytol.* **55**, 45–9.

BURNETT, J. H. (1956 a). The mating system of fungi. I. *New Phytol.* **55**, 50–90.

BUSTON, H. W., JABBAR, A. and ETHERIDGE, D. E. (1953). The influence of hexose phosphates, calcium and jute extract on the formation of perithecia by *Chaetomium globosum. J. gen. Microbiol.* **8**, 302–6.

Buston, H. W. and Khan, A. H. (1956). The influence of certain micro-organisms on the formation of perithecia by *Chaetomium globosum*. *J. gen. Microbiol.*, in the press.

Byrde, R. J. W. (1953). Observations on the sporulation of *Sclerotinia fructigena* on mummified apples and plums in late spring and summer. *Rep. agric. hort. Res. Sta. Bristol*, 1953, pp. 163–6.

Cantino, E. C. (1951). Metabolism and morphogenesis in a new *Blastocladiella*. *Leeuwenhoek ned Tijdschr.* **17**, 59–96.

Cantino, E. C. (1952). The biochemical nature of morphogenetic patterns in *Blastocladiella*. *Amer. Nat.* **86**, 399–404.

Cantino, E. C. (1953). The role of metabolism and α-ketoglutarate oxidase in the growth and differentiation of the aquatic phycomycete *Blastocladiella emersonii*. *Trans. N.Y. Acad. Sci.*, Series II, **15**, 159–63.

Cantino, E. C. (1954). The role of metabolism in the morphological differentiation of the water fungus *Blastocladiella emersonii*. Preliminary programme, Int. Bot. Congr., Paris.

Cantino, E. C. (1956). The relation between cellular metabolism and morphogenesis in *Blastocladia*. *Mycologia*, **48**, 225–40.

Cantino, E. C. and Horenstein, E. A. (1954). Cytoplasmic exchange without gametic copulation in the water mold *Blastocladiella emersonii*. *Amer. Nat.* **88**, 143–54.

Cantino, E. C. and Hyatt, M. T. (1953). Phenotypic 'sex' determination in the life history of a new species of *Blastocladiella emersonii*. *Leeuwenhoek ned Tijdschr.* **19**, 25–70.

Cantino, E. C. and Hyatt, M. T. (1953a). Carotenoids and oxidative enzymes in the aquatic phycomycetes *Blastocladiella* and *Rhizophlyctis*. *Amer. J. Bot.* **40**, 688–94.

Castle, E. S. (1928–9). Dark adaptation and light-growth response of *Phycomyces*. *J. gen. Physiol.* **12**, 391–400.

Castle, E. S. (1929–30). Phototropism and the light-sensitive system of *Phycomyces*. *J. gen. Physiol.* **13**, 421–35.

Castle, E. S. (1930–1). The phototropic sensitivity of *Phycomyces* as related to wave-length. *J. gen. Physiol.* **14**, 701–11.

Castle, E. S. (1931). Phototropic 'indifference' and the light-sensitive system of *Phycomyces*. *Bot. Gaz.* **91**, 207–12.

Castle, E. S. (1931–2). On 'reversal' of phototropism in *Phycomyces*. *J. gen. Physiol.* **15**, 487–9.

Castle, E. S. (1932–3). Dark adaptation and the dark growth response of *Phycomyces*. *J. gen. Physiol.* **16**, 75–88.

Castle, E. S. (1936). The origin of spiral growth in *Phycomyces*. *J. cell. comp. Physiol.* **8**, 493–502.

Castle, E. S. (1937). The distribution of velocities of elongation and of twist in the growth zone of *Phycomyces* in relation to spiral growth. *J. cell. comp. Physiol.* **9**, 477–89.

Castle, E. S. (1938). Orientation of structure in the cell wall of *Phycomyces*. *Protoplasma*, **31**, 331–45.

Castle, E. S. (1942). Spiral growth and reversal of spiralling in *Phycomyces* and their bearing on primary wall structure. *Amer. J. Bot.* **29**, 664–72.

CASTLE, E. S. (1953). Problems of oriented growth and structure in *Phycomyces*. *Quart. Rev. Biol.* **28**, 364–72.

CHEAL, W. F. and DILLON-WESTON, W. A. R. (1938). Observations on pear scab (*Venturia pirina* Aderh.). *Ann. appl. Biol.* **25**, 206–8.

CHODAT, R. and SCHOPFER, W. H. (1927). Carotine et sexualité. *C.R. Soc. Phys. Hist. nat. Genéve*, **44**, 176–9.

CHRISTENBERRY, G. A. (1938). A study of the effect of light of various periods and wavelengths on the growth and asexual reproduction of *Choanephora cucurbitarum* (Berk. and Rev.) Thaxter. *J. Elisha Mitchell sci. Soc.* **54**, 297–310.

CLAUSSEN, P. (1912). Zur Entwicklungsgeschichte der Ascomyceten (*Pyronema confluens*). *Z. Bot.* **4**, 1–64.

COHEN, M. (1950). Direct observation of the formation of sexual bodies by combination of hyphae of two *Phytophthora* isolates. *Phytopathology*, **40**, 5–6.

COKER, W. C. (1923). *The Saprolegniaceae*. Chapel-Hill, N.C.

COONS, G. H. (1916). Factors involved in the growth and pycnidium formation of *Plenodomus fuscomaculans*. *J. agric. Res.* **5**, 713–69.

CORNER, E. J. H. (1929). Studies in the morphology of Discomycetes. I. The marginal growth of apothecia. *Trans. Brit. mycol. Soc.* **14**, 263–75.

CORNER, E. J. H. (1929a). Studies in the morphology of Discomycetes. II. The structure and development of the ascocarp. *Trans. Brit. mycol. Soc.* **14**, 275–91.

CORNER, E. J. H. (1932). A *Fomes* with two systems of hyphae. *Trans. Brit. mycol. Soc.* **17**, 51–81.

CORNER, E. J. H. (1932a). The fruit-body of *Polystictus xanthopus*. *Ann. Bot., Lond.*, **46**, 71–111.

CORNER, E. J. H. (1934). An evolutionary study in agarics: *Collybia apalosarca* and the veils. *Trans. Brit. mycol. Soc.* **19**, 39–88.

CORNER, E. J. H. (1947). Variation in the size and shape of spores, basidia and cystidia in Basidiomycetes. *New Phytol.* **46**, 195–228.

CORNER, E. J. H. (1948). *Asterodon*, a clue to the morphology of fungus fruit-bodies: with notes on *Asteromella* and *Asterostromella*. *Trans. Brit. mycol. Soc.* **31**, 234–65.

CORNER, E. J. H. (1950). '*Clavaria*' *and allied genera*. Oxford.

COUCH, J. N. (1926). Heterothallism in *Dictyuchus*, a genus of the water molds. *Ann. Bot., Lond.*, **40**, 848–81.

CRAIGIE, J. H. (1927). Experiments on sex in rust fungi. *Nature, Lond.*, **120**, 116–17.

CRAIGIE, J. H. (1927a). Discovery of the function of the pycnidia of the rust fungi. *Nature, Lond.*, **120**, 765–7.

CRAIGIE, J. H. (1931). An experimental investigation of sex in the rust fungi. *Phytopathology*, **21**, 1001–40.

CURZI, M. (1932). Studi su lo '*Sclerotium rolfsii*'. *Boll. Staz. Pat. veg. Roma*, N.S., **11**, 306–73.

CURZI, M. (1932a). Contributo alla conoscenza della biologia e della sistematica degli stipiti dello '*Sclerotium rolfsii*'. *R.C. Accad. Lincei*, **15**, Ser. 6, 241–5.

DADE, H. A. (1937). New Gold Coast fungi, I. *Trans. Brit. mycol. Soc.* **21**, 16–28.

DAS GUPTA, S. N. (1933). Formation of pycnidia in *Cytosporina ludibunda* by the intermingling of two infertile strains. *Ann. Bot., Lond.*, **47**, 689–90.

DE BARY, A. (1881). Untersuchungen über *Peronospora* und Saprolegnieen. *Beitr. morph. Physiol. Pilze*, **4**.

DÉMÉTRIADÈS, S. D. (1953). Études sur la biologie du *Sclerotinia sclerotiorum* (Lib.) Massée. III. L'action du magnésium et du soufre sur le développement du champignon et la formation de ses sclérotes. IV. L'utilisation de diverses sources d'azote. *Ann. Inst. phytopath. Benaki*, **7**, 15–20, 27–35. Abstract in *Rev. appl. Mycol.* **34**, 740–1 (1955).

DENNY, F. E. (1933). Oxygen requirements of *Neurospora sitophila* for formation of perithecia and growth of mycelium. *Contr. Boyce Thompson Inst.* **5**, 95–102.

DERX, H. G. (1926). Heterothallism in the genus *Penicillium*. Preliminary note. *Trans. Brit. mycol. Soc.* **11**, 108–12.

DODGE, B. O. (1920). The life history of *Ascobolus magnificus*. *Mycologia*, **12**, 115–34.

DODGE, B. O. (1927). Nuclear phenomena associated with heterothallism and homothallism in the ascomycete *Neurospora*. *J. agric. Res.* **35**, 289–305.

DODGE, B. O. (1931). Heterothallism and hypothetical hormones in *Neurospora*. *Bull. Torrey bot. Cl.* **58**, 517–22.

DOUDOROFF, M., KAPLAN, N. and HASSID, W. Z. (1943). Phosphorolysis and synthesis of sucrose with a bacterial preparation. *J. biol. Chem.* **151**, 351–61.

DRAYTON, F. L. (1932). The sexual function of the microconidia in certain Discomycetes. *Mycologia*, **24**, 345–8.

DRIVER, C. H. and WHEELER, H. E. (1955). Asexual hormone in *Glomerella*. *Mycologia*, **47**, 311–16.

DUDDINGTON, C. L. (1955). Fungi that attack microscopic animals. *Bot. Rev.* **21**, 377–439.

EGERTON, C. W. (1914). Plus and minus strains in *Glomerella*. *Amer. J. Bot.* **1**, 244–54.

EMERSON, R. (1950). Current trends of experimental research on the aquatic Phycomycetes. *Ann. Rev. Microbiol.* **4**, 169–200.

EMERSON, R. and CANTINO, E. C. (1948). The isolation, growth and metabolism of *Blastocladia* in pure culture. *Amer. J. Bot.* **35**, 157–71.

EMERSON, R. and FOX, D. L. (1940). γ-Carotene in the sexual phase of the aquatic fungus, *Allomyces*. *Proc. Roy. Soc. B*, **128**, 275–93.

EMMONS, C. W. (1935). The ascocarps in species of *Penicillium*. *Mycologia*, **27**, 128–50.

ERGLE, D. R. and BLANK, L. M. (1947). A chemical study of the mycelium and sclerotia of *Phymatotrichum omnivorum*. *Phytopathology*, **37**, 153–61.

ERRERA, L. (1884). Die grosse Wachstumsperiode beiden Frechtträgern von *Phycomyces*. *Bot. Ztg.* **42**, 498–503.

EZEKIEL, W. N., TAUBENHAUS, J. T. and FUDGE, J. F. (1934). Nutritional requirements of the root-rot fungus, *Phymatotrichum omnivorum*. *Plant Physiol.* **9**, 187–216.

FISCHER, F. C. and WERNER, G. (1955). Eine Analyse des Chemotropismus einiger Pilze, insbesondere der Saprolegniaceen. *Hoppe-Seyl. Z.* **300**, 211–36.

FOSTER, J. W. and WAKSMAN, S. A. (1939). Fumaric acid formation associated with sexuality in a strain of *Rhizopus nigricans. Science,* **89**, 37.

FREY-WYSSLING, A. (1952). *Deformation and Flow in Biological Systems,* pp. 236–40. Amsterdam.

GADD, C. H. (1925). *Phytophthora faberi. Ann. R. bot. Gdns, Peradeniya,* **9**, 47–89.

GALLEMAERTS, V. (1910). De la zonation des cultures de champignons en boîte de Pétri. *Rec. Inst. bot. 'Leo. Errara', Brux.,* **8**, 213–23.

GALLOWAY, L. D. (1936). Report of the Imperial Mycologist. *Sci. Rep. agric. Inst. Pusa, 1934–5,* pp. 120–30.

GARRETT, S. D. (1949). A study of violet root rot. II. Effect of substratum on survival of *Helicobasidium purpureum* colonies in the soil. *Trans. Brit. mycol. Soc.* **32**, 217–23.

GARRETT, S. D. (1953). Rhizomorph behaviour in *Armillaria mellea* (Vahl). Quél. I. *Ann. Bot. N.S.* **17**, 63–79.

GARTON, G. A., GOODWIN, T. W. and LIJINSKY, W. (1950). The biogenesis of β-carotene in the fungus *Phycomyces blakesleeanus. Biochem. J.* **46**, 35–40.

GARTON, G. A., GOODWIN, T. W. and LIJINSKY, W. (1951). Studies in carotenogenesis. I. General conditions governing β-carotene synthesis by the fungus *Phycomyces blakesleeanus. Biochem. J.* **48**, 154–63.

GLEN-BOTT, J. I. (1955). On *Helicodendron tubulosum* and some similar species. *Trans. Brit. mycol. Soc.* **38**, 17–30.

GOIDANICH, G. and MEZETTI, A. (1948). Ricerche sulla biologia della *Melanospora damnosa* (Sacc.) Lindau. II. Contributo ricerche fisiologiche. *Ann. Sper. agr.,* N.S., **2**, 489–514.

GOLDRING, D. (1936). The effect of environment upon the production of sporangia and sporangiola in *Blakeslea trispora* Thaxter. *Ann. Mo. bot. Gdn.* **23**, 527–41.

GOODWIN, T. W. (1950). Carotenoids and reproduction. *Biol. Rev.* **25**, 391–431.

GOODWIN, T. W. (1952). Fungal carotenoids. *Bot. Rev.* **18**, 291–316.

GOTTLIEB, D. (1950). The physiology of spore germination in fungi. *Bot. Rev.* **16**, 229–57.

GRÄNTZ, F. (1898). Ueber den Einfluss des Lichtes auf die Entwickelung einiger Pilze. Inaug. Dissert. Leipzig.

GRASER, M. (1919). Untersuchungen über das Wachstum und Reizbarkeit der Sporangienträger von *Phycomyces nitens. Bot. Zbl.* **36**, 414–93.

GREEN, E. (1930). Observations on certain Ascobolaceae. *Trans. Brit. mycol. Soc.* **15**, 321–2.

GREGORY, P. H. (1939). The life history of *Ramularia vallisumbrosae* Cav. on *Narcissus. Trans. Brit. mycol. Soc.* **23**, 24–54.

GREGORY, P. H. (1952). Presidential Address. Fungus spores. *Trans. Brit. mycol. Soc.* **35**, 1–18.

GWYNNE-VAUGHAN, H. C. I. (1928). Sex and nutrition in fungi. *Brit. Assoc. Rept. 96th Meeting,* pp. 185–99.

HAFIZ, A. (1951). Cultural studies on *Ascochyta rabiei* with special reference to zonation. *Trans. Brit. mycol. Soc.* **34**, 259–69.

HALL, M. P. (1933). An analysis of the factors controlling the growth form of certain fungi, with special reference to *Sclerotinia (Monilia) fructigena*. *Ann. Bot., Lond.*, **47**, 538–78.

HAMILTON, R. I. and BOOSALIS, M. G. (1955). Asexual reproduction in *Cephalothecium gregalium*. *Phytopathology*, **45**, 293–4.

HANSEN, E. (1883). Recherches sur la physiologie et la morphologie des ferments alcooliques, II. *C.R. Lab. Carlsberg*, **2**, 29–102.

HANSEN, H. N. (1938). The dual phenomenon in imperfect fungi. *Mycologia*, **30**, 442–55.

HANSEN, H. N. and SMITH, R. E. (1932). The mechanism of variation in imperfect fungi: *Botrytis cinerea*. *Phytopathology*, **30**, 953–64.

HARDER, R. (1927). Zur Fräge nach der Rolle von Kern und Protoplasma in Zellgeschehen und bei der Übertragung von Eigenschaften. *Z. Bot.* **19**, 337–407.

HARDER, R. and SÖRGEL, G. (1938). Über einen neuen plano-isogamen Phycomyceten mit Generationswechsel und seine phylogenetische Bedeutung. *Nachr. Ges. Wiss. Göttingen*, **3**, 119–27.

HARTER, L. L. (1939). Influence of light on the length of the conidia in certain species of *Fusarium*. *Amer. J. Bot.* **26**, 234–43.

HASHIOKA, Y. (1938). Relation of temperature and humidity to *Sphaerotheca fuliginea* (Schlecht.) Poll. with special reference to germination, viability and infection. *Trans. nat. Hist. Soc. Formosa*, **27**, 129–45.

HATCH, W. R. (1935). Gametogenesis in *Allomyces arbuscula*. *Ann. Bot., Lond.*, **49**, 623–49.

HAWKER, L. E. (1936). The effect of certain accessory substances on the sporulation of *Melanospora destruens* and of some other fungi. *Ann. Bot., Lond.*, **50**, 699–718.

HAWKER, L. E. (1938). Effect of growth substances on growth and fruiting of *Melanospora destruens*. *Nature, Lond.*, **142**, 1038.

HAWKER, L. E. (1939). The influence of various sources of carbon on the formation of perithecia by *Melanospora destruens* Shear in the presence of accessory growth factors. *Ann. Bot., Lond.*, N.S., **3**, 455–68.

HAWKER, L. E. (1939a). The nature of the accessory growth substances influencing growth and fruiting of *Melanospora destruens* Shear and of some other fungi. *Ann. Bot., Lond.*, N.S., **3**, 657–76.

HAWKER, L. E. (1942). The effect of vitamin B_1 on the concentration of glucose optimal for the fruiting of certain fungi. *Ann. Bot., Lond.*, N.S., **6**, 631–6.

HAWKER, L. E. (1944). The effect of vitamin B_1 on the utilization of glucose by *Melanospora destruens* Shear. *Ann. Bot., Lond.*, N.S., **8**, 79–90.

HAWKER, L. E. (1947). Effect of temperature of storage on the rate of loss of fertility of stock cultures of *Melanospora destruens*. *Nature, Lond.*, **159**, 136.

HAWKER, L. E. (1947a). Further experiments on growth and fruiting of *Melanospora destruens* Shear in the presence of various carbohydrates, with special reference to the effects of glucose and sucrose. *Ann. Bot., Lond.*, N.S., **11**, 245–59.

HAWKER, L. E. (1948). Stimulation of the formation of perithecia of *Melanospora destruens* Shear by small quantities of certain phosphoric esters of glucose and fructose. *Ann. Bot., Lond.*, N.S., **12**, 77–9.

HAWKER, L. E. (1950). *Physiology of Fungi.* London.

HAWKER, L. E. (1951). Morphological and physiological studies on *Sordaria destruens* (Shear) comb. nov. (syn. *Melanospora destruens*), *Sordaria fimicola* and *Melanospora zamii. Trans. Brit. mycol. Soc.* **34**, 174–86.

HAWKER, L. E. (1954). British hypogeous fungi. *Phil. Trans.* B, **237**, 429–546.

HAWKER, L. E. (1955). Hypogeous fungi. *Biol. Rev.* **30**, 127–58.

HAWKER, L. E. and CHAUDHURI, S. D. (1946). Growth and fruiting of certain ascomycetous fungi as influenced by the nature and concentration of carbohydrate in the medium. *Ann. Bot., Lond.,* N.S., **10**, 185–94.

HEALD, F. D. and POOL, V. W. (1908). The influence of chemical stimulation upon the production of perithecia of *Melanospora pampeane* Speg. *Rep. Neb. agric. Exp. Sta., 96th Meeting,* pp. 185–96.

HEDGECOCK, G. C. (1906). Zonation in artificial cultures of *Cephalothecium* and other fungi. *Ann. Mo. bot. Gdn,* **17**, 115–17.

HEMMONS, L. M., PONTECORVO, G. and BUFTON, A. W. J. (1953). Perithecium analysis and relative heterothallism, in Pontecorvo, G. 'The genetics of *Aspergillus nidulans'. Advances in Genetics,* **5**, 194–201.

HENRIKSSON, L. E. and MORGAN-JONES, J. F. (1951). The effect of temperature, pH and malt extract upon growth and perithecial development of two *Gnomonia* species. *Svensk bot. Tidskr.* **45**, 648–56.

HENRY, B. W. and ANDERSEN, A. L. (1948). Sporulation by *Piricularia oryzae. Phytopathology,* **38**, 265–78.

HEYN, A. N. J. (1936). Further investigations on the mechanism of cell elongation and the properties of the cell wall in connection with elongation. IV. Investigations on the molecular structure of chitin cell wall of sporangiophores of *Phycomyces* and its probable bearing on the phenomenon of spiral growth. *Protoplasma,* **25**, 372–96.

HEYN, A. N. J. (1939). Some remarks on the mechanism of spiral growth of the sporangiophores of *Phycomyces* and a suggestion for its further exploration. *Proc. Acad. Sci. Amst.* **42**, 431–7.

HIGGINS, B. B. (1927). Physiology and parasitism of *Sclerotium rolfsii. Phytopathology,* **17**, 417–48.

HIRSCH, H. M. (1954). Environmental factors influencing the differentiation of protoperithecia and their relation to tyrosinase and melanin formation in *Neurospora crassa. Physiol. Plant.* **7**, 72–97.

HIRST, J. M. (1955). The early history of a potato blight epidemic. *Plant Pathology,* **4**, 44–50.

HOPP, H. (1938). Sporophore formation by *Fomes applanatus* in culture. *Phytopathology,* **28**, 356–8.

HORNE, A. S. and MITTER, J. H. (1927). Studies in the genus *Fusarium.* V. Factors determining septation and other features in the section Discolor. *Ann. Bot., Lond.,* **41**, 519–47.

HOUSTON, B. S. and OSWALD, J. W. (1946). The effect of light and temperature on conidium production by *Helminthosporium gramineum* in culture. *Phytopathology,* **36**, 1049–55.

HUGHES, S. J. (1953). Conidiophores, conidia and classification. *Canad. J. Bot.* **31**, 577–659.

HUMPHREY, J. E. (1892). The Saprolegniaceae of the United States. *Trans. Amer. phil. Soc.* **17**, 63–148.

INGOLD, C. T. (1939). *Spore Discharge in Land Plants.* Oxford.

INGOLD, C. T. (1942). Aquatic Hyphomycetes of decaying alder leaves. *Trans. Brit. mycol. Soc.* **25**, 339–417.

INGOLD, C. T. (1946). Size and form in Agarics. *Trans. Brit. mycol. Soc.* **29**, 108–13.

INGOLD, C. T. (1953). *Dispersal in Fungi.* Oxford.

INGOLD, C. T. (1954). Ascospore form. *Trans. Brit. mycol. Soc.* **37**, 19–21.

INGOLD, C. T. (1954a). Presidential Address. Fungi and water. *Trans. Brit. mycol. Soc.* **37**, 97–107.

INGOLD, C. T. and Cox, V. J. (1955). Periodicity of spore discharge in *Daldinia. Ann. Bot., Lond.,* N.S., **19**, 201–10.

INGRAM, M. (1955). *An Introduction to the Biology of Yeasts.* London.

JINKS, J. L. (1954). Somatic selection in fungi. *Nature, Lond.,* **174**, 409–10.

KANOUSE, B. B. (1932). A physiological and morphological study of *Saprolegnia parasitica. Mycologia,* **24**, 431–51.

KAUFFMANN, C. H. (1908). A contribution to the physiology of the Saprolegniaceae, with special reference to the variation of the sexual organs. *Ann. Bot., Lond.,* **22**, 361–87.

KEHL, H. (1937). Ein Beitrag zur Morphologie und Physiologie der Zygophoren von *Mucor mucedo. Arch. Mikrobiol.* **8**, 379–406.

KEYWORTH, W. G. (1942). The occurrence in artificial culture of tremelloid outgrowths on the pilei of *Coprinus ephemerus. Trans. Brit. mycol. Soc.* **25**, 307–10.

KLEBS, G. (1898, 1899, 1900). Zur Physiologie der Fortpflanzung einiger Pilze. *J. wiss. Bot.* **32**, 1–70; **33**, 513–97; **35**, 80–203.

KOHLER, F. (1935). Beiträg zur Kenntnis der Sexualreaktionen von *Mucor mucedo. Planta,* **23**, 358–78.

KOUYEAS, V. (1953). On the sexuality of *Phytophthora parasitica* Dastur. *Ann. Inst. phytopath. Benaki,* **7**, 40–53.

KRAFCZYK, H. (1931). Die Zygosporenbildung bei *Pilobolus crystallinus. Ber. dtsch. bot. Ges.* **49**, 141–6.

KRAFCZYK, H. (1935). Die Bildung und Keimung der Zygosporen von *Pilobolus crystallinus* und sein heterokaryotisches Mycel. *Beitr. Biol. Pfl.* **23**, 349–96.

KREUTZER, W. A., BODINE, E. W. and DURRELL, L. W. (1940). A sexual phenomenon exhibited by certain isolates of *Phytophthora capsici. Phytopathology,* **30**, 951–7.

KREUTZER, W. A., DURRELL, L. W. and BODINE, E. W. (1940). Pathogenicity and sexual phenomena exhibited by *Phytophthora capsici. J. Colo.-Wyo. Acad. Sci.* **2**, 35. (Abstract in *Biol. Abstr.* **14**, 1208.)

LAMBERT, E. B. (1933). Effect of excess carbon dioxide on growing mushrooms. *J. agric. Res.* **47**, 599–608.

LARGE, E. C. (1953). Potato blight forecasting investigation in England and Wales, 1950–2. *Plant Pathology,* **2**, 1–15.

LEAVER, F. W., LEAL, J. and BREWER, C. R. (1947). Nutritional studies on *Piricularia oryzae. J. Bact.* **54**, 401–8.

LENDNER, A. (1897). Des influences combinées de la lumière et du substratum sur le développement des champignons. *Ann. Sci. nat. Bot.* **3**, 1–64.

LENDNER, A. (1918). Les mucorinées géophiles recoltées à Bourg-Saint-Pierre. *Bull. Soc. bot. Genève*, **10**, 362–76.

LEONIAN, L. H. (1924). A study of factors promoting pycnidium-formation in some Sphaeropsidales. *Amer. J. Bot.* **9**, 19–50.

LEONIAN, L. H. (1931). Heterothallism in *Phytophthora*. *Phytopathology*, **21**, 941–55.

LEONIAN, L. H. (1935). The effect of auxins on *Phytophthora cactorum*. *J. agric. Res.* **51**, 277–86.

LEONIAN, L. H. (1936). Effects of auxins from some green algae upon *Phytophthora cactorum*. *Bot. Gaz.* **47**, 854–9.

LEONIAN, L. H. and LILLY, V. G. (1940). Studies on the nutrition of fungi. V. Factors affecting zygospore formation in *Phycomyces blakesleeanus*. *Amer. J. Bot.* **27**, 670–5.

LESTER-SMITH, W. C. (1927). Some observations on the oospores of *Phytophthora*. *Ann. R. bot. Gdns, Peradeniya*, **9**, 243–57.

LEUPOLD, U. (1950). Die Vererbung von Homothallie und Heterothallie bei *Schizosaccharomyces* Bombe. *C.R. Lab. Carlsberg*, Sér. physiol. **24**, 381–480.

LEVI, J. D. (1956). Mating reactions in yeast. *Nature, Lond.*, **177**, 753–4.

LEVINE, S. and ORDAL, Z. J. (1946). Factors influencing the morphology of *Blastomyces dermatitidis*. *J. Bact.* **52**, 687–94.

LILLY, V. G. and BARNETT, H. L. (1947). The influence of pH and certain growth factors on mycelial growth and perithecial formation by *Sordaria fimicola*. *Amer. J. Bot.* **34**, 131–8.

LILLY, V. G. and BARNETT, H. L. (1948). Growth rates, vitamin deficiencies and sclerotia formation by some Sclerotiniaceae. *Proc. W. Va Acad. Sci.* **20**, 69–74.

LILLY, V. G. and BARNETT, H. L. (1949). The influence of concentrations of nutrients, thiamin, and biotin upon growth and formation of perithecia and ascospores by *Chaetomium convolutum*. *Mycologia*, **41**, 186–96.

LILLY, V. G. and BARNETT, H. L. (1951). *Physiology of the Fungi*. New York.

LINDEGREN, C. C. (1934). The genetics of *Neurospora*. V. Self-sterile bisexual heterokaryons. *J. Genetics*, **28**, 425–35.

LINDEGREN, C. C. (1936). Heterokaryosis and hormones in *Neurospora*. *Amer. Nat.* **70**, 404–6.

LOCKWOOD, L. B. (1937). Hydrogen ion concentration and ascus formation. *Mycologia*, **29**, 289–90.

LODDER, J. (1934). *Die anaskosporogen Hefen*. Amsterdam.

LOPRIORE, G. (1895). Über die Einwirkung der Kohlensäure auf das Protoplasma der lebenden Pflanzenzelle. *J. wiss. Bot.* **28**, 531–626.

MADELIN, M. F. (1956). Studies on the nutrition of *Coprinus lagopus* Fr., especially as affecting fruiting. *Ann. Bot., Lond.*, N.S. **20**, 307–30.

MADER, E. O. (1943). Some factors inhibiting the fructification and production of the cultivated mushroom, *Agaricus campestris* L. *Phytopathology*, **33**, 1134–45.

MAURIZIO, A. (1899). Beiträge zur Biologie der Saprolegnieen. *Z. Fisch.* **7**, 655.

McCALLAN, S. E. A. and CHAN, S. Y. (1944). Inducing sporulation of *Alternaria solani* in culture. *Contr. Boyce Thompson Inst.* **13**, 323-36.

McCORMICK, F. A. (1925). Perithecia of *Thielavia bascicola* Zopf. in culture and the stimulation of their production by extracts from other fungi. *Bull. Conn. agric. Exp. Sta.* no. 269, pp. 539-54.

McVICKAR, D. L. (1942). The light-controlled diurnal rhythm of asexual reproduction of *Pilobolus*. *Amer. J. Bot.* **29**, 372-80.

MILLER, J. J., CALVIN, J. and TREMAINE, J. H. (1955). A study of factors influencing sporulation of *Saccharomyces cerevisiae*. *Canad. J. Microbiol.* **1**, 560-73.

MOHENDRA, K. R. (1928). A study of the changes undergone by certain fungi in artificial culture. *Ann. Bot., Lond.*, **42**, 863-89.

MOHENDRA, K. R. and MITRA, M. (1930). On the cultural behaviour of *Sphaeropsis malorum* Pk. *Ann. Bot., Lond.*, **44**, 541-55.

MOLLIARD, M. (1903). Rôle des bactéries dans la production des périthèces des *Ascobolus*. *C.R. Acad. Sci., Paris*, **136**, 899-901.

MOREAU, F. and MOREAU, C. (1938). La formation hormonale des périthèces chez les *Neurosporas*. *C.R. Acad. Sci., Paris*, **206**, 369-70.

MOREAU, F. and MORUZI, C. (1931). Recherches expérimentales sur la formation des périthèces chez les *Neurosporas*. *C.R. Acad. Sci., Paris*, **192**, 1475-8.

MRAK, E. M. and BONAR, L. (1938). The effect of temperature on asci and ascospores in the genus *Debaryomyces*. *Mycologia*, **30**, 182-6.

MULDER, E. G. (1938). Influence of copper on growth of microorganisms. *Ann. Ferment.* **4**, 513-33.

NAKATA, K. (1929). Studies of *Sclerotium rolfsii*. Part VII. The results of successive cultures and selections within pure lines of the fungus. *Bull. Sci. Fak. Terk. Kjusu Univ.* **3**, 292-9. (Abstract in *Rev. appl. Myc.* **9**, 613.)

NARASHIMAN, M. J. (1930). Studies in the genus *Phytophthora* in Mysore. I. Heterothallic strains of *Phytophthora*. *Phytopathology*, **20**, 201-14.

NICKERSON, W. J. (1954). Experimental control of morphogenesis in microorganisms. *Ann. N.Y. Acad. Sci.* **60**, 50-7.

NICKERSON, W. J. and MANKOWSKI, Z. (1953). Role of nutrition in the maintenance of the yeast-shape in *Candida*. *Amer. J. Bot.* **40**, 584-92.

NICKERSON, W. J. and THIMANN, K. V. (1941). The chemical control of conjugation in *Zygosaccharomyces*. *Amer. J. Bot.* **28**, 617-21.

NICKERSON, W. J. and THIMANN, K. V. (1943). The chemical control of conjugation in *Zygosaccharomyces*, II. *Amer. J. Bot.* **30**, 94-101.

NIETHAMMER, A. (1938). Wachstumversuche mit mikroskopischen Boden Pilzen. *Arch. Mikrobiol.* **9**, 23-30.

NITIMARGI, N. M. (1937). Studies in the genera *Cytosporina, Phomopsis* and *Diaporthe*. VII. Chemical factors influencing sporing characters. *Ann. Bot., Lond.*, **49**, 19-40.

OGILVIE, L. (1944). Downy mildew of lettuce. *Rep. agric. hort. Res. Sta., Bristol, 1943*, pp. 90-4.

OLSON, E. O. (1950). Genetics of *Ceratostomella*. I. Strains in *Ceratostomella fimbriata* (Ell. and Hals) Elliot from sweet potatoes. *Phytopathology*, **39**, 548-61.

OORT, A. J. P. (1930). Die Sexualität von *Coprinus fimetarius*. *Rec. Trav. bot. néerl.* **27**, 85–148.

OORT, A. J. P. (1931). The spiral growth of *Phycomyces*. *Proc. Acad. Sci. Amst.* **34**, 564–75.

OORT, A. J. P. and ROELOFSEN, P. A. (1932). Spiralwachstum, Wandbau und Plasmaströmung bei *Phycomyces. Proc. Acad. Sci. Amst.* **35**, 898–908.

ORTH, H. (1937). Der Einfluss der Luftfeuchtigkeit auf das Keimverhalten der Sporangien von *Phytophthora infestans* (Mont.) de Bary, des Energens der Kartoffelfäule. *Z. Pfl. Krankh.* **47**, 425–47.

PAGE, R. M. (1956). Studies on the development of asexual reproductive structures in *Pilobolus. Mycologia*, **48**, 206–24.

PAPE, H. (1937). Beiträge zur Biologie und Bekampfung des Kleekrebses (*Sclerotinia trifoliorum* Erikss.). *Arb. biol. Abt. (Anst. Reichsanst.), Berl.*, **22**, 159–247.

PARR, R. (1918). The response of *Pilobolus* to light. *Ann. Bot., Lond.*, **32**, 177–205.

PASTEUR, L. (1876). *Études sur la bière.* Paris.

PAUL, W. R. C. (1929). A comparative morphological and physiological study of a number of strains of *Botrytis cinerea*, with special reference to their virulence. *Trans. Brit. mycol. Soc.* **14**, 118–34.

PEIRIS, J. W. L. (1947). The *Botrytis* disease of *Gladiolus*, together with a physiological study of certain *Botrytis* species. Thesis for the degree of Ph.D. in the University of London.

PERKINS, S. J. (1952). The effect of low temperature on the formation of zygospores of some members of the Mucorales. Thesis in part fulfilment of the regulations for the degree of B.Sc. in the University of Bristol.

PFEFFER, W. (1903). *Physiology of Plants*, vol. II, p. 90. Oxford.

PLUNKETT, B. E. (1953). Nutritional and other aspects of fruit body production in pure cultures of *Collybia velutipes* (Curt.) Fr. *Ann. Bot., Lond.*, N.S., **17**, 193–216.

PLUNKETT, B. E. (1956). The influence of factors of the aeration complex and light upon fruit-body form in pure cultures of an Agaric and a Polypore. *Ann. Bot., Lond.*, N.S. (in the press).

PONTECORVO, G. and GEMMELL, A. R. (1944). Genetic proof of heterokaryosis in *Penicillium notatum. Nature, Lond.*, **154**, 514–16.

POP, L. J. J. (1938). Protoplasmic streaming in relation to spiral growth of *Phycomyces. Proc. Acad. Sci. Amst.* **41**, 661–72.

PORGES, N. (1932). Chemical composition of *Aspergillus niger* as modified by zinc sulphates. *Bot. Gaz.* **94**, 197–207.

PRESTON, R. D. (1952). *The Molecular Architecture of Plant Cell Walls*, pp. 182–201. New York.

RANDS, R. D. (1917). The production of spores by *Alternaria solani* in pure culture. *Phytopathology*, **7**, 316–17.

RAPER, J. R. (1936). Heterothallism and sterility in *Achlya* and observations on the cytology of *A. bisexualis. J. Elisha Mitchell sci. Soc.* **52**, 274–89.

RAPER, J. R. (1939). Role of hormones in the sexual reaction of heterothallic *Achlyas. Science*, **89**, 321–2.

RAPER, J. R. (1939a). Sexual hormones in *Achlya*. I. Indicative evidence for a hormonal coordinating mechanism. *Amer. J. Bot.* **26**, 639–50.

RAPER, J. R. (1940). Sexual hormones in *Achlya*. II. Distance reactions, conclusive evidence for a hormonal coordinating mechanism. *Amer. J. Bot.* **27**, 162–73.

RAPER, J. R. (1940a). Sexuality in *Achlya ambisexualis*. *Mycologia*, **32**, 710–27.

RAPER, J. R. (1942). Sexual hormones in *Achlya*. III. Hormone A and the initial male reaction. *Amer. J. Bot.* **29**, 159–66.

RAPER, J. R. (1942a). Sexual hormones in *Achlya*. V. Hormone A, a male secreted augmentor or activator of hormone A. *Proc. Nat. Acad. Sci.*, *Wash.*, **28**, 509–15.

RAPER, J. R. (1950). Sexual hormones in *Achlya*. VI. The hormones of the A complex. *Proc. Nat. Acad. Sci.*, *Wash.*, **36**, 524–33.

RAPER, J. R. (1950a). Sexual hormones in *Achlya*. VII. The hormonal mechanism in homothallic species. *Bot. Gaz.* **112**, 1–24.

RAPER, J. R. (1951). Sexual hormones in *Achlya*. *Amer. Sci.* **39**, 110–20.

RAPER, J. R. (1952). Chemical regulation of sexual processes in the thallophytes. *Bot. Rev.* **18**, 447–545.

RAPER, J. R. (1953). Tetrapolar sexuality. *Quart. Rev. Biol.* **28**, 233–57.

RAPER, J. R. and HAAGEN-SMIT, A. J. (1942). Sexual hormones in *Achlya*. IV. Properties of hormone A of *A. bisexualis*. *J. biol. Chem.* **143**, 311–20.

RAPER, K. B. and FENNELL, D. L. (1952). Homothallism and heterothallism in the *Penicillium luteum* series. *Mycologia*, **44**, 101–11.

RAULIN, G. J. (1869). Études chimique sur la végétation. *Ann. Sci. nat. Bot.* 5 sér. **9**, 93–299.

REAVER, M. J. (1954). Effect of certain chloronitrobenzenes on germination, growth and sporulation of some fungi. *Ann. appl. Biol.* **41**, 448–60.

REISCHER, H. S. (1949). The effect of temperature on the papillation of oogonia of *Achlya colorata*. *Mycologia*, **41**, 398–402.

ROBBINS, W. J. (1939). Growth substances and gametic reproduction in *Phycomyces*. *Bot. Gaz.* **101**, 428–99.

ROBBINS, W. J. (1939a). Growth substances in agar. *Amer. J. Bot.* **26**, 772–8.

ROBBINS, W. J. (1940). Effects of extracts of *Phycomyces* on its development. *Amer. J. Bot.* **27**, 559–64.

ROBBINS, W. J. (1941). Further observations on Factor Z. *Bot. Gaz.* **102**, 520–35.

ROBBINS, W. J. (1943). Specificity of hypoxanthine for *Phycomyces*. *Proc. Nat. Acad. Sci.*, *Wash.*, **29**, 201–2.

ROBBINS, W. J. and HAMNER, K. C. (1940). Effect of potato extracts on growth of *Phycomyces*. *Bot. Gaz.* **101**, 912–27.

ROBBINS, W. J. and KAVANAGH, F. (1938). Vitamin B_1 or its intermediates and the growth of certain fungi. *Amer. J. Bot.* **25**, 229–36.

ROBBINS, W. J. and KAVANAGH, F. (1938a). Thiamin and growth of *Pythium butleri*. *Bull. Torrey bot. Cl.* **65**, 453–61.

ROBBINS, W. J. and KAVANAGH, F. (1942). Guanine and Factor Z, growth substances for *Phycomyces*. *Proc. Nat. Acad. Sci.*, *Wash.*, **28**, 4–8.

ROBBINS, W. J. and KAVANAGH, F. (1942a). Hypoxanthine, a growth factor of *Phycomyces*. *Proc. Nat. Acad. Sci.*, *Wash.*, **28**, 65–9.

ROBBINS, W. J. and SCHMIDT, M. B. (1945). Factor Z_2 and gametic reproduction of *Phycomyces*. *Amer. J. Bot.* **32**, 320–6.

ROBERG, M. (1928). Über Wirkung von Eisen-, Zink- und Kupfer-salzen für *Aspergillus*. *Z. Bakt.* 2, **76**, 333–71.

ROBERTS, P. (1954). The effect of low temperatures on zygospore production in *Mucor sexualis*. Thesis in part fulfilment of the regulations for the degree of B.Sc. in the University of Bristol.

ROBERTSON, N. and MACFARLANE, I. (1946). The occurrence of perithecia of the oak mildew in Britain. *Trans. Brit. mycol. Soc.* **29**, 219–20.

ROBINSON, W. (1925). On some features of growth and reproduction in *Sporodinia grandis*, Link. *Trans. Brit. mycol. Soc.* **10**, 307–14.

ROBINSON, W. (1926). The conditions of growth and development of *Pyronema confluens* Tul. *P. omphaloides* (Bull.) Fuckel. *Ann. Bot., Lond.*, **40**, 245–72.

ROELOFSEN, P. A. (1950). The origin of spiral growth in *Phycomyces* sporangiophores. *Rec. Trav. bot. néerl.* **42**, 72–110.

ROELOFSEN, P. A. (1950a). Cell-wall structure in the growth-zone of *Phycomyces* sporangiophores. I. Model experiments and microscopical observations. *Biochim. biophys. Acta*, **6**, 340–56.

ROELOFSEN, P. A. (1951). Cell-wall structure in the growth-zone of *Phycomyces* sporangiophores. II. Double retraction and electron microscopy. *Biochim. biophys. Acta*, **6**, 357–73.

ROGERS, C. H. (1939). Relation of moisture and temperature to growth of the cotton root-rot fungus. *J. agric. Res.* **58**, 701–9.

RONSDORF, L. (1931). Über die chemischen Bedingungen von Wachstum und Zygotenbildung bei *Phycomyces blakesleeanus*. *Planta*, **14**, 482–514.

SARTORY, A. (1912). Sporulation d'une levure sans l'influence d'une bactérie. *C.R. Soc. biol., Paris*, **72**, 582–60.

SARTORY, A. (1916). De l'influence d'une bactérie sur la production des périthèces chez un *Aspergillus*. *C.R. Soc. biol., Paris*, **79**, 174–5.

SARTORY, A. (1918). Sporulation par symbiose chez des champignons inférieurs. *C.R. Acad. Sci., Paris*, **167**, 302–5.

SATINA, S. and BLAKESLEE, A. F. (1925). Studies on biochemical differences between the (+) and (−) sexes in *Mucors*. I. Tellurium salts as indicators of the reduction reaction. *Proc. Nat. Acad. Sci., Wash.*, **11**, 528–34.

SATINA, S. and BLAKESLEE, A. F. (1926). Studies on biochemical differences between the (+) and (−) sexes in *Mucors*. II. A preliminary report on the Manilov reactor and other tests. *Proc. Nat. Acad. Sci., Wash.*, **12**, 191–6.

SATINA, S. and BLAKESLEE, A. F. (1926a). The *Mucor* parasite *Parasitella* in relation to sex. *Proc. Nat. Acad. Sci., Wash.*, **12**, 202–7.

SCHERR, G. H. and WEAVER, R. H. (1953). The dimorphism phenomenon in yeast. *Bact. Rev.* **17**, 51–92.

SCHOPFER, W. H. (1931). Recherches expérimentales sur la formation des zygotes chez *Phycomyces blakesleeanus*. L'influence des substances vitaminiques. *Bull. Soc. bot. Suisse*, **40**, 87–111; **41**, 73–95.

SCHOPFER, W. H. (1931a). Études de l'influence des extraits de levures et des concentrées de vitamines B sur la sexualité d'un champignon. *C.R. Soc. Phys. Hist. nat. Genève*, **48**, 105–7.

SCHOPFER, W. H. (1937). La spécificité d'action de l'aneurine sur *Phycomyces*. *Bull. Soc. bot. Suisse*, **47**, 460-4.

SCHOPFER, W. H. (1943). *Plants and Vitamins*. New York.

SCHOPFER, W. H. and BLUMER, S. (1940). Le pouvoir de synthèse d'un facteur de croissance par *Schizophyllum commune* (haplontes et diplontes). *Protoplasma*, **34**, 524-9.

SHEAR, C. L. and DODGE, B. O. (1927). Life histories and heterothallism of the red bread mould fungi of the *Monilia* group. *J. agric. Res.* **34**, 1019-42.

SNYDER, W. C. and HANSEN, H. N. (1941). The effect of light on the taxonomic characters in *Fusarium*. *Mycologia*, **33**; 580-91.

SØRGEL, G. (1953). Über den Entwicklungsgang von *Mycosphaerella pinodes* (Berk. and Blox.) Stone. II. Die Einfluss der Wasserstoffionenkonzentration auf die Ausbildung der Fortpflanzungsorgane. *Arch. Mikrobiol.* **19**, 372-97.

STAMPS, D. J. (1953). Oospore production in paired cultures of *Phytophthora* species. *Trans. Brit. mycol. Soc.* **36**, 255-8.

STEINBERG, R. D. (1919). A study of some factors in the chemical stimulation of the growth of *Aspergillus niger*. *Amer. J. Bot.* **6**, 37-48.

STEVENS, F. L. (1928). Effects of ultra-violet radiation on various fungi. *Bot. Gaz.* **86**, 210-25.

STEVENS, F. L. and HALL, J. G. (1909). Variation of fungi due to environment. *Bot. Gaz.* **48**, 1-30.

SUBRAMANIAN, C. V. and PAI, K. V. S. (1953). Relation of nitrogen to growth and sporulation of *Fusarium vasinfectum*. *Proc. Indian Acad. Sci.*, sect. B, **37**, 149-57.

SWEET, H. R. (1942). Studies on the biology of two species of *Magnusia*. II. Effect of humidity on conidial germination, growth and reproduction. *Amer. J. Bot.* **29**, 436-41.

THREN, R. (1937). Gewinnung und Kultur von monokaryotischen und dikaryotischen Mycel. Ein Beitrag zur Physiologie und Genetik des Gerstenflugbrandes (*Ustilago nuda* (Jens) Kellerm. et Sw.). *Z. Bot.* **31**, 337-91.

THUNG, T. H. (1926). Opmerkingen over *Peronospora parasitica* op Kool (Bemerkungen über die Kohl-*Peronospora*). *Tijdschr. PlZiekt.* **32**, 161-79.

TIMNICK, M. B., LILLY, V. G. and BARNETT, H. L. (1951). The influence of light and other factors upon sporulation of *Diaporthe phaseolorum* var. *batatitis* from soya bean. *Phytopathology*, **41**, 327-36.

TOWNSEND, B. B. (1952). The morphology and physiology of the sclerotia and rhizomorphs of certain fungi. Thesis for the degree of Ph.D. in the University of Bristol.

TOWNSEND, B. B. and WILLETTS, H. J. (1954). The development of sclerotia of certain fungi. *Trans. Brit. mycol. Soc.* **37**, 213-21.

TURIAN, G. (1952). Carotenoides et différentiation sexuelle chez *Allomyces*. *Experientia*, **8**, 302.

TURIAN, G. and HAXO, F. T. (1954). Minor polygene components in the sexual phase of *Allomyces javanicus*. *Bot. Gaz.* **115**, 254-60.

VAN BEVERWIJK, A. L. (1951). Zalewski's '*Clathrosphaera spirifera*'. *Trans. Brit. mycol. Soc.* **34**, 280-90.

VANDENDRIES, R. (1934). Les barrages sexuels chez *Lenzites betulina* (L.), Fr. *C.R. Acad. Sci., Paris,* **198**, 193–5.

VANDENDRIES, R. and BRODIE, H. J. (1933). Nouvelles investigations dans la domaine de la sexualité des Basidiomycètes et étude expérimentelle des barrages sexuels. *La Cellule,* **42**, 165–209.

VENKAT RAM, C. S. (1952). Soil bacteria and chlamydospore formation in *Fusarium solani. Nature, Lond.,* **170**, 889.

VERKAIK, C. (1930). Ueber das Enstehen von Zygophoren von *Mucor mucedo* (+) unter Beeinflussung eines von *Mucor mucedo* (−) abgescheidenen Stoffes. *Proc. Acad. Sci. Amst.* **33**, 656–8.

WALTER, H. (1924). Plasmaquellung und Wachstum. *Z. Bot.* **13**, 673–718.

WATERHOUSE, G. M. (1930). The production of conidia in the genus *Phytophthora. Trans. Brit. mycol. Soc.* **15**, 311–21.

WESTERGAARD, M. and HIRSCH, H. (1954). Environmental and genetic control of differentiation in *Neurospora. Symp. Colston res. Soc. Bristol,* 1954.

WESTERGAARD, M. and MITCHELL, H. K. (1947). *Neurospora.* V. A synthetic medium favouring sexual reproduction. *Amer. J. Bot.* **34**, 573–7.

WHITEHOUSE, H. L. K. (1949). Heterothallism and sex in fungi. *Biol. Rev.* **24**, 411–47.

WHITEHOUSE, H. L. K. (1949a). Multiple-allelomorph heterothallism in the fungi. *New Phytol.* **48**, 212–44.

WHITEHOUSE, H. L. K. (1951). A survey of heterothallism in the Ustilaginales. *Trans. Brit. mycol. Soc.* **34**, 340–55.

WIESNER, J. (1873). Ueber den Einfluss der Temperatur auf die Entwickelung von *Penicillium glaucum. Akad. wiss. Wien. Abth. Math.* **17**, 835–6.

WILKINS, W. H. and HARRIS, G. C. M. (1946). The ecology of the larger fungi. V. *Ann. appl. Biol.* **33**, 179–87.

WILKINS, W. H. and PATRICK, S. H. M. (1940). The ecology of the larger fungi. *Ann. appl. Biol.* **27**, 15–34.

WILSON, E. E. (1927). Effects of fungus extracts upon the initiation and growth of perithecia of *Venturia inaequalis* in pure culture. *Phytopathology,* **17**, 835–6.

YARWOOD, C. E. (1936). The diurnal cycle of the powdery mildew *Erysiphe polygoni. J. agric. Res.* **52**, 645–57.

YARWOOD, C. E. (1937). The relation of light to the diurnal cycle of sporulation of certain downy mildews. *J. agric. Res.* **54**, 365–73.

YARWOOD, C. E. (1941). Diurnal cycle of ascus maturation of *Taphrina deformans. Amer. J. Bot.* **28**, 355–7.

YARWOOD, C. E. (1943). Onion downy mildew. *Hilgardia,* **14**, 595–691.

ZENTMYER, G. A. (1952). A substance stimulating sexual reproduction in *Phytophthora cinnamomi.* (Abstract in *Phytopathology,* **42**, 24.)

ZICKLER, H. (1943). Cited by Raper (1952) from M. Hartmann, *Die Sexualität,* pp. 385–6.

INDEX

Absidia
  glauca, 84
  spinosa, 78
Acaulium nigrum, 64
Achlya, 86–9, 95
  ambisexualis, 71, 86
  bisexualis, 86
  colorata, 29
aconitase, 74
adenine, 29, 64, 65, 71
aeration, effect on sclerotia, 20
  effect on sporulation, 33–5
  of soil, 103
algae, as source of growth-substance, 85
Allomyces, 96, 97, 98
  cystogena, 96
  javanicum, 96
  moniliformis, 96
Alternaria solani, 37
amino acids, 61, 88
ammonia, 33, 34, 60
ammonium salts, 21, 60
ammonium sulphate, 18, 69
ammonium tartrate, 57
anastomosis of hyphae, 77, 93
antheridia, 76, 85, 86, 87–9, 93, 95
antheridial hyphae, 86–9
antibiotic substances, 19, 37
antirrhinum rust, 105
apothecia, 38, 40, 44, 58, 78
arabinose, 53, 55
arabo-furanose, 55
Arcangeliella, 10
Armillaria mellea, 23
Arthrobotrys oligospora, 50
asci, 26, 29, 90
  'crossed', 94
Ascobolus, 33, 70
  magnificus, 78, 93
Ascochyta rabiei, 25
ascogenous hyphae, 76
ascogonia, 41, 76
ascospores, 25, 42, 56, 59, 90, 92, 99
asparagine, 21, 51, 57, 60
aspartic acid, 21
Aspergillus, 16, 17, 27, 46
  fumigatus, 56
  nidulans, 94, 99
  niger, 63, 64, 70, 90
Asterodon, 14
auxin, 85
α-ketoglutarate, 74
  oxidase, 97

Barbituric acid, 88
barium, 62
'barrage sexuel', 94
barriers to growth, 38
bicarbonate, 74
biotin (vitamin H), 29, 64, 65, 67, 68, 69, 70, 90
Blakeslea trispora, 52
Blastocladia pringsheimiana, 34
Blastocladiella, 98
  emersonii, 74, 96, 97
  variabilis, 96
Blastomyces dermatitidis, 17
Bombardia lunatus, 93
Botrytis, 23, 34, 73
  allii, 15, 19
  cinerea, 15, 19, 20, 21, 32, 34, 60
Bremia lactucae, 30
brown rot of fruits, 104
Bulgaria inquinans, 10

Calcium, 37, 62
  chloride, 62
Candida albicans, 17
carbohydrate, 21, 22, 23, 52–7
  concentration of, 54 (Fig. 3)
carbon, 18, 20, 21, 22, 52–7
  dioxide, 34, 35
carbon-nitrogen ratio, 21–3, 57–61
carotene, 44, 95, 96, 98
carotenoids, 44, 63, 95, 96
casein, 21
  hydrolysate, 58
Cephalothecium
  gregalium, 52
  roseum, 39
Ceratostomella fimbriata, 27, 69, 71, 91
Chaetomium
  convolutum, 69
  globosum, 36, 37, 38, 56, 62
chemotropism, 88
  of trichogyne, 93
chlamydospores, 18–19, 36
Choanephora, 44
  cucurbitarum, 25, 29, 31, 34, 39, 41, 45, 71, 95
chromatography, paper, 56
Cladosporium fulvum, 105
clamp connexions, 93
Clathrosphaerina, 34
Clavaria spp., 9
Claviceps, 9
cleistocarps, 25, 26, 36, 91, 94
co-carboxylase, 67

Collybia velutipes, 32, 35, 36, 51, 57, 60, 70
concentration, of nutrients, 49–52
conidia, 25, 26, 30, 31, 36, 39, 40, 44, 50, 64, 99
  shape and septation, 60
conjugation, of Mucorales, 80–4
  of yeast, 70, 90
contact, stimulus of, 38, 88, 103
'contaminant', effect on sporulation, 66
copper, 63
Coprinus, 40, 43, 57
  congregatus, 57
  ephemerus, 32, 70
  lagopus, 49
  sassii, 60
  sterquilinus, 41
cultures, mixed, 65
  monospore, 77
cysteine, 18, 21
cystine, 21
Cystogenes, 96
Cytosporina, 60
  ludibunda, 95
cytoplasmic exchange of swarmers, 97
cytoplasmic inheritance, 99

Daldinia concentrica, 40
dark adaptation, 43
Debaryomyces spp., 29
decarboxylation, oxidative, 74
Diaporthe, 60
  phaseolorum, 42
dicarboxylic acids, 88
dicaryon, 76
Dictyuchus monosporus, 86
diphenylamine, 96
diploidization, 78

Egg albumin, 21
Elaphomyces spp., 9
Elaphomycetaceae, 104
Endomyces, 2, 10, 16
Endomycopsis capsularis, 17
enzyme systems, growth-substances as components of, 64
  trace elements as components of, 63
Erysiphe polygoni, 40
esters, phosphate, 55, 56
Euallomyces, 96
Eurotium, 10, 27
  herbariorum, 25, 36
  repens, 26
  rugulosum, 36

Factor Z, 70
fertile strains of *Melanospora destruens*, 99
fertilization, hormonal control of, 88
filtrates from paired cultures, 85
flavin, absorption of light by, 43, 44
Fomes, 4
  levigatus, 14

forecasting of disease epidemics, 104
fructose, 52, 53
  furanose, 55
fructose diphosphate, 55
fructose-1:6-diphosphate, 56
fruit-bodies, 4, 12–14
  abnormal, 32, 35, 43
  of agarics, 57
  dwarf, 57
  fleshy, 31
  subterranean, 103
  woody, 31
  xeromorphic, 32
fruit bodies, production of, in natural habitat, 102
fumaric acid, 83
fungi, hypogeous, 103
fungicides, 34, 37
Fusarium spp., 18, 34, 39, 43, 60, 61
  coeruleum, 34
  culmorum, 90
  discolor sulphureum, 44
  fructigenum, 25, 37, 56, 61
  moniliforme, 65
  oxysporum, 64
  solani, 19
  vasinfectum, 61

Galactose, 53
gallium, 63
gametangia, 81, 96
gametangial initials, 80
'gamma' factors, 97
Ganoderma, 4
  applanatum, 31
gelatine, 60
gemmae, 19
Genea spherica, 8
Geoglossum, 9
geographical races, 79
Glomerella cingulata, 92
glucose, 18, 21, 25, 51, 52, 53, 54, 55, 56, 57, 70
  concentration of, 22
glucose-1-phosphate, 55, 56
glutamic acid, 21
glutaric acid, 70, 88, 90
glycine, 21
glycocoll, 21
glycogen, 55
Gnomonia, 36
  erythrostroma, 50
  intermedia, 26
  vulgaris, 26
gravity, 45
growth-substances, 23, 57, 64–71
guanine, 64, 70, 71

Helicobasidium purpureum, 22
Helicodendron, 34
Heliscus aquaticus, 8
Helminthosporium gramineum, 45

heterocaryosis, 77
heterothallism, 77–9, 80, 81, 89, 91
  bipolar, 78
  incomplete, 92
  relative, 94
  tetrapolar, 78
hexose phosphates, 55, 62, 67
hexose sugars, 53
histamine, 80
hormone A, A^1, A^2, A^3, 88, 89
hormone B, 88, 89
hormone C, 88, 89
hormone D, 88, 89
hormones, inactivation of, 91
  in Mucorales, 83
  sex, 79–95
humidity, 20, 30–3
  interaction with light, 44
hybridization, imperfect, 78
  in *Phytophthora*, 85
hydantoin, 88
Hydnangium, 10
  carneum var. *xanthosporum*, 8
Hydnobolites cerebriformis, 8
hydrogen peroxide (as oxidizing agent), 45
hydrogen-ion concentration, 27, 39, 51, 68
  effect on sclerotia, 20
  effect on sporulation, 35–7
  of soil, 103
Hymenogaster, 10
  luteus, 104
  olivaceus, 104
  tener, 8, 104
hypogeous fungi, 38, 103
hypoxanthine, 64, 70, 71
Hysterangium, 10
  nephriticum, 8

Inhibiting substances, 23
inhibitor, 58
injury, mechanical, 37–8
inositol, 67, 71
interfertility between strains, 79
inulin, 49, 53, 55
invertase, 55
iron, 63
isogametes, 96

Jute extract, 56
  calcium in, 62

Lactose, 53, 55
Lemonniera aquatica, 8
lentil extract, 66
Lentinus lepideus, 43
leucine, 21
light, blue, 44
  effect on sclerotia, 20
  effect on sporulation, 38
  red-yellow, 44
  ultra-violet, 37, 44

lithium chloride, 63
lysine, 21

Magnesium, 61
  sulphate, 61
Magnusia, 31
malonic acid, 88
maltose, 53, 55, 70
manganese, 63
mannitol, 55
melanin, 73, 96
Melanogaster variegatus, 18
Melanospora, 56, 66
  damnosa, 90
  destruens, 25, 35, 37, 38, 43, 50, 52, 53, 54, 55, 56, 57, 66, 67, 68, 69, 71, 72, 90, 99
  pampeane, 65
metabolism, 71–5
methionine, 21
Microsphaera quercina, 105
mildew, downy, 30, 40, 50
  powdery, 31
mitochondria, 97
molybdenum, 63, 64
Monilia, see *Sclerotinia*
Monilia fructigena, 104
morphogenesis, 71
Mucor, 2, 16, 17, 39, 46, 71, 72, 73, 83, 95
  flavidus, 39
  hiemalis, 32, 78, 80, 81
  mucedo, 34, 80, 82
  parasitella, 84
  racemosus, 18, 26
  sexualis, 25, 26, 28, 71
multiple allelomorphs, 79
mummied fruits, 104
mushroom, 14, 35
mutants of *Neurospora*, 61
mycelium, exhaustion of, 103
mycorrhizal fungi, 103
Mycosphaerella pinodes 36

Natural habitat, reproduction in, 100–6
Neurospora, 56, 61, 63, 78, 91, 98
  crassa, 28, 58, 59, 73
  sitophila, 33, 90, 93
nitrogen, source of, 18, 21, 22, 57–61
nuclei, pairing of, 76
nucleic acid, 71
nutrients, concentration of, 49–52
nutrition, effect on sclerotia, 20–3
  effect on sporulation, 48–75

Oak mildew, 105
oatmeal agar, 31
oidia, 16–18
oogonia, 31, 85–9, 95
oogonial initials, 86, 87, 89
oospheres, differentiation of, 89

oospores, 26, 71, 85, 89
  hybrid, 85
oxygen, 33, 35
  consumption of, 81

Paramecium, 97
parasites on fungi, 84
  on plants, 17, 104–6
pathogens, animal, 17
Penicillium, 32
  javanicum, 36
  luteum, 91
peptone, 21, 22, 60
periodicity, diurnal, 40
perithecia, 25, 26, 27, 28, 35, 36, 38,
  42, 50, 51, 52, 53, 55, 56, 57, 59,
  68, 90, 91, 99
Peronospora
  destructor, 30
  parasitica, 29
Pezizales, 9
pH, *see* hydrogen-ion concentration
Phomopsis, 60
phosphates, 61–2, 95
phosphoglyceric acid, 56
phosphorus, 61–2
phosphorylation of carbohydrates, 55
Phycomyces, 42, 44, 73
  blakesleeanus, 11, 12, 27, 32, 37, 52,
    58, 60, 70, 72, 80, 81, 95
  nitens, 32, 80
Phyllosticta sp., 39
Phymatotrichum omnivorum, 19, 23
phytofluene, 96
Phytophthora, 31, 71, 89
  arecae, 84
  cactorum, 31
  capsici, 85
  cinnamomi, 85
  colocasiae, 85
  cryptogea, 85
  drechsleri, 85
  faberi, 84
  infestans, 30, 104
  meadii, 85
  omnivora, 84
  palmivora, 85
  parasitica, 84, 85
pigmentation of spores, 63
Pilobolus, 40, 43, 44
  crystallinus, 80
  microsporus, 26, 43
pimelic acid, 88
Piricularia oryzae, 34, 71
planozygotes, 96
plant extracts, 60
plant pathogens, 104–6
Plectascales, 9
Plenodomus fuscomaculans, 33, 45
polyphenol oxidase, 63
polyenes, 96

Polyporellus brumalis, 32, 35, 44; 46
polypores, 50, 105
Polyporus
  brumalis, 33, 35
  squamosus, 43
Polystictus xanthopus, 13
potassium, 61
  nitrate, 21, 58, 61
potato blight, 104
predacious fungi, 50
protoperithecia, 28, 58
Pseudomonas saccharophila, 55
pseudomycelium, 16
Pseudoperonospora humuli, 29
pseudothecia, 36
Puccinia
  antirrhini, 105
  graminis, 6
purines, 29, 64, 70, 71
pycnidia, 33, 36, 45, 95, 98
pyridoxine, 64, 65
pyrimidine, 22, 67, 68
Pyronema confluens, 37, 38, 40, 44, 49, 54,
  58
pyruvic acid, 69
  decarboxylation of, 67
Pythium, 30, 31, 71

Raffinose, 53, 55
rainfall, effect of, 102, 103
Ramularia vallisumbrosae, 32
reduction division, 77
reproduction, by spores, 3–5
  vegetative, 2
respiration, 46, 56, 67, 81
riboflavin, 70, 90
Rhizoctonia, 23
  solani, 15, 19, 20, 22
rhizomorphs, 23
Rhizophlyctis, 97
Rhizopus nigricans, 64, 80, 83
rhythmic response, 40

Saccharomyces cerevisiae, 26, 56, 90
Saccobolus depauperatus, 70
Saprolegnia, 49, 86
  ferax, 88
  mixta, 26
Sapromyces reinschii, 86
Schizophyllum commune, 70, 94
Schizosaccharomyces pombe, 91
sclerotia, 19–23
  development of, 14–15
sclerotial production, 60
Sclerotinia, 23
  fructicola, 39
  fructigena, 39
  gladioli, 15, 19, 78
  sclerotiorum, 21
  trifoliorum, 21

127

Sclerotium
  cepivorum, 15
  rolfsii, 15, 19, 20, 21, 22
Sepedonium chrysogenum, 18
sex, determination of, 95–8
  physiology of, 76–99
sex hormones, 79–95
sexual anomalies, 89
sexuality, loss of, 98–99
smuts, 18
  heterothallism in, 79
soil conditions, effect of, 103
soil moisture, 103
  effect on sclerotia, 20
soil temperature, 103
Sordaria destruens, 25; and see *Melanospora*
  fimicola, 35, 68, 72
Sphaerographium fraxini, 45
Sphaeropsidales, 49
Sphaeropsis malorum, 98
Sphaerotheca humuli var. *fuliginea*, 31
sporangia, 11, 12, 25, 26, 30, 39, 44, 52, 105
  of *Phycomyces*, 11–12
  resting, 34, 74, 96
sporangioles, 44, 52
sporangiophores, effect of light on, 42–4
  of *Phycomyces*, 11
spore discharge, 6
spore dispersal, 45
  type, effect of concentration on, 52
spore-bearing structures, growth of, 7–15
spores, dwarf, 92
  functions of, 5–6
  growth of, 7–11
Sporodinia grandis, 25, 26, 27, 32, 52
sprout mycelium, 16
staling factors, 66
starch, 18, 53, 55
Stephanospora, 10
sterile mutants of *Melanospora destruens*, 99
stimulus, air-borne, 82
strains, complementary, 79
  plus and *minus*, 78, 80, 82, 83, 94
  self-fertile, 84, 92
  self-sterile, 92
strontium, 62
sucrose, 25, 51, 53, 54, 55, 56, 57
  inversion of, 55
sugar, hexose, 18, 55
  reducing, 18
sulphites, 61
sulphur, 61–2
swarmers, 97
Syzygites, see *Sporodinia*

Taphrina, 17, 18
  deformans, 40

telemorphosis, 80
teleutospores, 105
temperature, effect of in natural habitat, 102, 103
  effect on sclerotia, 19
  effect on sporulation, 24–9
Thamnidium elegans, 44, 52
thiamin, 22, 29, 64, 65, 67, 68, 69, 70, 71, 90
  components of, 22, 67
thiazole, 22, 43, 67
thigmatropism, 89
Thraustotheca, 89
  primoachlya, 86
tomato mould, 105
toxins, effect on sporulation, 37
trace elements, 63–4
translocation, 38, 46
Trichoderma viride, 34, 62
trichogyne, 76, 93
truffles, 10, 103
tryptophane, 21
tyrosinase, 73
tyrosine, 21
Tuber, 9, 10
Tuberales, 104

Urea, 21, 61
uredospores, 105
Ustilago maydis, 29
  nuda, 94

Valine, 21
vegetative reproduction, physiology of, 16–23
ventilation of glasshouses, 105
Venturia inaequalis, 50
  pyrina, 31, 50
vitamin B₁, see thiamin
vitamins, 64–71

Water, effect on sporulation, 30–3
water moulds, 19, 30, 90
  sex hormones of, 86–9

Yeast, baker's, 34
  utilization of sucrose by, 55

Zinc, 63, 64
zonation, 25, 39
zone of restraint, 80
zoosporangia, 26, 30, 34, 74, 96
zoospores, 3, 74
zygophore, 80, 82
Zygorhynchus, 71
Zygosaccharomyces, 90
  acidifaciens, 70
zygospores, 25, 26, 27, 28, 29, 32, 58, 60, 64, 70, 71, 78
zygotropism, 80

Printed in the United States
By Bookmasters